Nils Knauer

Signal Losses of Outdoor-Indoor Wave Propagation Paths

I0009668

Nils Knauer

Signal Losses of Outdoor-Indoor Wave Propagation Paths

Signal Fading through Modern Window Systems

VDM Verlag Dr. Müller

Bibliographic information by the German National Library: The German National Library lists this publication at the German National Bibliography; detailed bibliographic information is available on the Internet at http://dnb.d-nb.de.

This works including all its parts is protected by copyright. Any utilization falling outside the narrow scope of the German Copyright Act without prior consent of the publishing house is prohibited and may be subject to prosecution. This applies especially to duplication, translations, microfilming and storage and processing in electronic systems.

Any brand names and product names mentioned in this book are subject to trademark, brand or patent protection and are trademarks or registered trademarks of their respective holders. The use of brand names, product names, common names, trade names, product descriptions etc. even without a particular marking in this works is in no way to be construed to mean that such names may be regarded as unrestricted in respect of trademark and brand protection legislation and could thus be used by anyone.

Contents

1. Introduction

The invention of wireless systems was one of the greatest steps that mankind has made. The last twenty years, in particular, have shown how significant this technology is to normal, everyday life. The number of subscribers to wireless services, such as the mobile phone, increased in an impressive manner not comparable with any other technical development.

With the beginning of the 21st century, approximately 600 million people utilised mobile phones for calls and writing text messages. By the end of this first decade, several estimations have predicted, that every second person on earth, which means more than three billion people, will possess a mobile phone [1]. The performance of the mobile phone will also improve over that time. The subscribers are currently able to use their small, multifunctional handheld device for high data rate services, such as the Internet or watching TV. What new applications will the mobile support in a few years?

Data cables may disappear in almost every computer network. Computers may be linked together without any visible connections. Bluetooth, a short distance wireless technology, connects devices such as printers or scanners with computers and also Internet access is possible. This list could be easily extended, for instance by considering navigation systems via satellite.

Modern information technology is primarily based on mobility and high – performance systems, so that everybody is reachable by telephone and may also access all desired data or information at any place of the world. In other words, wireless technology is so common in our life that it is not conceivable to act without it anymore.

The number of services already existing, which are, in terms of the technical point of view, operating in an uncomplicated, low frequency range, is very big. But they are operating very close to maximum capacity. Thus, the adoption of new high data rate services requires a different frequency range, which will be above 10 GHz.

Signal coverage problems are still recognisable, manifesting themselves as noise or bad speech quality, for example, when making a mobile phone call in a hostile environment. These undesired effects will be much stronger when the operating frequency is increased. The use of more power to achieve better signal strengths is possible up to a certain limit, but the level of contamination by electromagnetic waves is already quite high and, for health reasons, to exceed these limits is, in any case, prohibited by law.

The analysis of electromagnetic wave propagation into buildings, under certain conditions and within certain constraints is the main thrust of the work described in this thesis. UHF signals, penetrating into buildings, are strongly attenuated by brick or concrete walls. The only exploitable signals are those transmitted through windows. Therefore, investigations will be made on signal transmission through various windows for frequencies, representing existing broadcast services, such as FM radio, and unicast services such as GSM and UMTS.

Chapter 2 deals with the general mathematical background of electromagnetism. First of all, a brief overview of the propagation effects is presented. Based on Maxwell's Equations, the simplest form of waves, i.e. the uniform plane wave, is presented. Referring to that wave model, the electric and magnetic field components of electromagnetic waves are introduced, as also are the most significant propagation constants.

Chapter 3 presents a description of glass, its history, the materials utilised for windowpanes, in the past as well as currently, and the setup of modern window constructions. Afterwards, a theoretical overview of the propagation of electromagnetic waves through a double-glazed and multiple-glazed window is presented. The definition of both linear polarisation types is displayed, and their equations to describe the electric and magnetic field components are introduced. A brief description of the ideas behind the software model created and the difficulties when conductive materials, for instance coatings, are components of a window, is presented.

Chapter 4 describes the measurement setup and procedures, the antennas utilised and the investigations performed over the frequency range. Finally, technical notes in terms of the measurements are presented, which are very important to consider.

Chapter 5 illustrates the results of the complete measurement series, while chapter 6 deals with their discussion, including whether inaccuracies could falsify them.

Chapter 7 presents the final conclusion of the whole work, for example the classification of the results in terms of the construction of windowpanes in the near future.

2. Electromagnetic propagation through solid material

Electrodynamics, which is based on *Maxwell's Equations*, describes inter alia the interaction between the *electric field* E and the *magnetic field* H. This interaction results in generating electromagnetic waves (EM waves). The physics of optics provides an adequate and simple description for all basic phenomena affecting EM waves during their transmission through objects [2]. The first section of this chapter presents an introduction of these basic propagation effects.

Section 2.2 deals with Maxwell's Equations, which provide the common mathematical formulas [3]. Also, the electromagnetic propagation of Uniform plane waves, through solid materials, is presented.

2.1. Propagation effects in solid material

The behaviour of EM waves is similar to the well-known propagation effects of light rays, when they impinge on an object. Thus, it is appropriate to start by mentioning the most common light propagation phenomena. Light is thus, initially, considered to travel in rays (geometrical optics) instead of waves. Afterwards, these easy physical laws can be extended to the more complex equations for EM waves, where information like magnitude and phase are considered (see Section 2.2). The most significant propagation effects are reflection, diffraction and scattering [4]. Due to the fact that EM waves may change their direction of propagation when they pass an object, it is also necessary to mention refraction too. Figure 2-1 illustrates possible phenomena when an EM wave propagates into another medium:

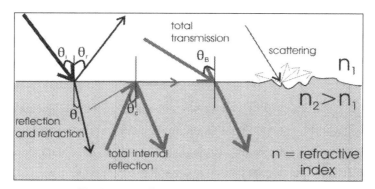

Fig. 2-1:   Several propagation phenomena

2.1.1 Reflection

We expect reflections when an EM wave impinges on an object, which is very large, compared to the wavelength. Smooth surfaces, e.g. a windowpane are objects, which cause a certain amount of energy to be reflected, while the rest will propagate through the interface between the two media [4]. The angle of incidence θ_i is equal to the angle of reflection θ_r, is the well-known *Law of Reflection* [5]. The definition of these angles is illustrated in Figure 2-1. Both rays and the normal to the plane surface lie in the same plane, which is called the *plane of incidence.*

2.1.2 Refraction

Refraction is when a light ray, not normally incident, changes its direction when entering from one medium into another one. The angle of refraction, which describes this changing of direction, depends on the so-called *refractive index n* of the medium [6]. With respect to Figure 2-1, the mathematical equation, developed by Snell, for this phenomenon may be written as:

$$\frac{\sin \theta_i}{\sin \theta_t} = \frac{n_2}{n_1} \qquad [2.1.1]$$

The *refractive index* n can be obtained with the relative permittivity ε_r of the concerning medium:

$$n = \sqrt{\varepsilon_r} \qquad [2.1.2]$$

The permittivity ε of a medium, as will be shown later on, is an important parameter to determine the individual electrical properties of materials in terms of their penetrability for electromagnetic fields [3]. The permittivity is defined as follows:

$$\varepsilon = \varepsilon_0 \cdot \varepsilon_r \qquad [2.1.3]$$

The permittivity, ε, is the product of the permittivity of vacuum, ε_0, and the relative permittivity, ε_r, of the medium.

The *permeability, μ,* represents the degree of magnetisation of a material. Therefore, this parameter is related to the magnetic field, and can be obtained from the product of the permeability of vacuum, μ_0, and the relative permeability of the medium, μ_r:

$$\mu = \mu_0 \cdot \mu_r \qquad [2.1.4]$$

The speed of light in vacuum, c_0, and the velocity, v_p, of the EM wave in a medium with respect to the material properties may be calculated with following equations:

$$c_0 = \frac{1}{\sqrt{\mu_0 \cdot \varepsilon_0}}$$

[2.1.5]

$$v_p = \frac{c_0}{\sqrt{\mu_r \cdot \varepsilon_r}} = \frac{1}{\sqrt{\mu \cdot \varepsilon}}$$

[2.1.6]

Not only the direction of propagation changes in the new medium, but also the phase velocity, v_P, and, deductively the wavelength λ. The criterion is the ratio of the refractive indices and can be written as:

$$\frac{\lambda_2}{\lambda_1} = \frac{v_{P2}}{v_{P1}} = \frac{n_1}{n_2}$$

[2.1.7]

It is also possible to express the velocity of a wave in a medium in terms of the speed of light in vacuum, c_0:

$$v_{P1} \cdot n_1 = v_{P2} \cdot n_2 = c_0$$

[2.1.8]

The incident vector, the refracted vector and the normal of the plane lie in the same plane, which is called the *plane of incidence* (see chapter 2.1.1).

2.1.2.1 Total internal reflection

When a light ray approaches the boundary of a medium whose refractive index n_2, is greater than 1, from within the medium and moving into air ($n_1=1$), the ray would refract away from the normal. At the so-called *critical angle*, θ_c, the refracted ray emanates perpendicular to the normal. Every incident ray, which is above the angle of θ_c, will be totally reflected back into the medium [5,7]. The equation for this special case can be written as:

$$n_2 \cdot \sin \theta_c = n_1$$

[2.1.9]

2.1.2.2 Total transmission

At another specific angle the light ray will be transmitted completely into the other medium, thus no reflection will occur. This angle is called the *Brewster Angle* θ_B, which is defined as:

$$\sin \theta_B = \sqrt{\frac{\varepsilon_1}{\varepsilon_1 + \varepsilon_2}}$$

[2.1.10]

ε_1 is the permittivity of the medium the light ray has left, while ε_2 refers to the permittivity of the medium the ray propagates.

2.1.3 Diffraction

If the direct path (line-of-sight; LOS) between the transmitter and the receiver is obstructed by objects, the EM waves bend around the obstacle. Based on Huygen's principle, which has explained that every point of a wavefront is a new point source of secondary waves, the signal can propagate in areas, which are not visible from the position of the transmitter [4]. These secondary waves originating at the edge of the obstacle's surface is illustrated in Figure 2-2. Diffraction depends not only on the geometry of the obstacle, it is also affected by the amplitude, phase and polarisation of the incident wave.

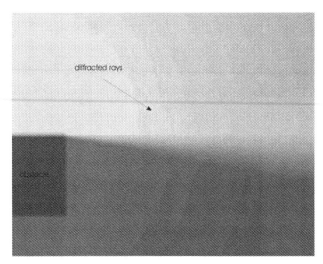

Fig. 2-2:   Diffraction at the edge of an obstacle

2.1.4 Scattering

Scattering effects can be observed, when a propagating wave impinges on an object whose size is less than the order of a wavelength. Examples of objects, which cause scattering, are rough surfaces, foliage or office supplies [4, 1].

A scattered ray would reradiate in any directions, which is very time-consuming to analyse or compute. In terms of this thesis scattering effects are negligible, due to the smooth surface and size windowpanes usually have.

2.1.5 Multi-path propagation

In almost every transmission of wireless communication systems, multi-path propagation occurs. The propagation is affected by reflection, diffraction and scattering on objects. The conclusion of the influence of all three phenomena is, that the EM waves propagate in multiple paths, e.g. the direct path (LOS) or paths determined by the propagation effects already mentioned. Due to superposition it is possible that signals cancel each other at a certain position, which results in no field strength and is called a dead spot. In an office building, moving obstacles, such as people, cause many field strength fluctuations. No LOS path exists, thus propagation effects constantly influence the signals. Additionally, due to the different path lengths, time delay spread will occur, which will limit the maximum symbol rate of transmission and thus the performance of a wireless system. Therefore, a reliable prediction of the field strength in an indoor environment is one of the most challenging and potentially rewarding areas of propagation investigation [1].

2.2. Electromagnetic Theory

For a more detailed investigation, electromagnetic theory has to be used. *Maxwell's Equations* are the basis of the explanation of the electromagnetic phenomena. It is possible to describe the generating of electric and magnetic fields due to current flows and charges for the static and time-variant cases. These equations arose out of several experiments by famous scientists such as Faraday, Ampere, Gauss, Lenz, Coulomb, Volta and others during the nineteenth century [3]. This chapter also presents an introduction of the uniform plane waves, a trivial form of EM wave, which can be described by *Maxwell's Equations*.

2.2.1 Maxwell's Equations

In the first place, *Maxwell's Equations* can determine the electric and the magnetic field, in terms of magnitude and phase. An EM wave consists of two components, the vectors of an electric (*E*) and a magnetic field (*H*) component. The general *Maxwell's Equations*, valid for vacuum or for the static case can be written as:

$$\oint_c E \cdot dl = -\frac{d}{dt} \cdot \int_s B \cdot ds \qquad \nabla \times E = -\frac{\partial B}{\partial t} \qquad [2.2.1]$$

$$\oint_c H \cdot dl = \int_s J \cdot ds + \frac{d}{dt} \int_s D \cdot ds \quad \nabla \times H = J + \frac{\partial D}{\partial t} \qquad \text{[2.2.2]}$$

$$\oint_s D \cdot ds = \int_v \rho \, dv \qquad \nabla \cdot D = \rho \qquad \text{[2.2.3]}$$

$$\oint_s B \cdot ds = 0 \qquad \nabla \cdot B = 0 \qquad \text{[2.2.4]}$$

Equation 2.2.1 is called *Faraday's Law of Induction.* Equations 2.2.2 is known as Ampère's Law, the third and fourth are *Gauss's Laws* for the electric and magnetic fields [3,8]. Equation 2.2.1 is an expression for describing a time-varying magnetic field *H*, which is designated by the magnetic flux density *B*. With the penetration of *H* into a surface *s*, an electric field *E* is induced along a closed contour called *c*, which is responsible for forces on every electric charge, ρ, along that contour. *Ampére's Law* states that the magnetic field *H* over a closed contour *c* is equal to the enclosed current, where the current density is represented by *J*, and the time-varying electric flux density *D* is related to the displacement current. The last-mentioned, which also describes charge conservation, is significant for describing EM wave propagation. This term was added by Maxwell to *Ampére's Law*, to also describe the dynamics of charges and their interaction with time-varying fields. *Gauss's* Law says that the electric flux *D* through a closed surface is equal to the enclosed charge. Finally, the last equation, referring to the magnetic field *H*, describes that the magnetic flux density *B* over a closed surface s, is equal to zero.

Subsequently, for simplification purposes, an assumption is made that the media of our math model is homogeneous, isotropic, source-free, free of charges and time-invariant. For that case, the following equations are valid:

$$\rho = 0 \qquad \text{[2.2.5]}$$

$$J = \sigma \cdot E = 0 \qquad \text{[2.2.6]}$$

Finally, *Maxwell's Equations* can be rewritten, relating the following expressions

$$D = \varepsilon \cdot E \qquad \text{[2.2.7]}$$

$$B = \mu \cdot H \qquad \text{[2.2.8]}$$

to the form which matches with the assumptions made for the math model:

$$\oint_c E \cdot dl = -\mu \cdot \frac{d}{dt} \cdot \int_s H \cdot ds \qquad \nabla \times E = -\mu \cdot \frac{\partial H}{\partial t} \qquad [2.2.9]$$

$$\oint_c H \cdot dl = \varepsilon \cdot \frac{d}{dt} \cdot \int_s E \cdot ds \qquad \nabla \times H = \varepsilon \cdot \frac{\partial E}{\partial t} \qquad [2.2.10]$$

$$\oint_s E \cdot ds = 0 \qquad \nabla \cdot E = 0 \qquad [2.2.11]$$

$$\oint_s H \cdot ds = 0 \qquad \nabla \cdot H = 0 \qquad [2.2.12]$$

2.2.2 Uniform plane waves

A uniform plane wave, which is the simplest form of an EM wave and is also named a TEM wave (*transverse electromagnetic mode*), is defined as a wave, where the vectors of E and H are perpendicular to each other, as well as to the direction of propagation. Figure 2-3 shows this uniform plane wave.

2.2.3 Energy transport

Fluctuating electromagnetic fields cause a power and energy flux in the direction the wave is propagating. The power density is represented by the so-called *Poynting vector S*, and is only valid in the far-field region of the transmitter. The direction of S is perpendicular to E and to H and is defined as:

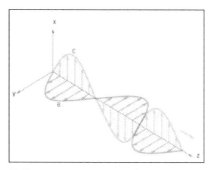

Fig 2-3: Uniform plane wave propagating in the z-direction

$$S = E \times H \qquad\qquad [2.2.13]$$

While, the time-average *Poynting vector* over one wavelength is:

$$S_{av} = \frac{1}{2} \operatorname{Re}(E \times H^*) \qquad\qquad [2.2.14]$$

The term $E \times H^*$ is generally complex, where the real part presents the power flux density and the imaginary part the reactive power [3].

Another possibility to calculate the energy transport is the radar equation. The power density at the receiver can be calculated with following equation:

$$S_R = \frac{P_T \cdot G_T}{4 \cdot \pi \cdot R^2} \qquad\qquad [2.2.15]$$

S_R is the power density at the receiver's position, while P_T is the energy radiated by the antenna and G_T the gain with respect to an isotropic radiator. R is the distance between transmitter and receiver [9]

2.2.4 Propagation constants and material properties

The properties of materials are significant when considering the penetration of EM waves through them. This work gives special attention to the energy losses of EM waves passing through objects. Based on the fact that we consider linear, homogeneous and isotropic media, the most important parameters will be presented in this chapter.

The parameters ε (*permittivity*), µ (*permeability*), and σ (*conductivity*) are constant in a homogeneous media, but they vary with frequency and thus they are designated as *dispersive*. To determine the losses of an EM wave, the *propagation constant* γ will be adopted. This parameter is defined as follows:

$$\gamma = \alpha + j \cdot \beta = \sqrt{j \cdot \omega\mu \cdot (\sigma + j \cdot \omega\varepsilon)} \qquad\qquad [2.2.16]$$

α is called the *attenuation constant*, and β is called as the *phase constant* or *wave number* [3]. These parameters are usually used when analysing wave propagation.

The general equations of α and β are defined as:

$$\alpha = \omega\sqrt{\mu\varepsilon} \cdot \sqrt{\frac{1}{2}\left[\sqrt{1+\left(\frac{\sigma}{\omega\varepsilon}\right)^2}-1\right]}$$ [2.2.17]

$$\beta = \omega\sqrt{\mu\varepsilon} \cdot \sqrt{\frac{1}{2}\left[\sqrt{1+\left(\frac{\sigma}{\omega\varepsilon}\right)^2}+1\right]}$$ [2.2.18]

Chapter 3 deals with the propagation of EM waves, especially in terms of the propagation factor, in more detail.

It is worth mentioning the anticipated *skin depth* δ of the wave [3]. The *skin depth* is defined as the depth an EM wave has travelled into a lossy medium when the signal falls to about $e^{-1} \triangleq 37$ % of its original strength. Skin depth can be obtained from the reciprocal of the *attenuation coefficient* in equation 2.2.17:

$$\delta = \frac{1}{\alpha}$$ [2.2.19]

It is important to note that the *skin depth* is distinctly higher than the thickness of the window coatings, usually utilised in modern window constructions, which are generally in the dimensions of micrometer or even in nanometres. But these coatings are high reflective, which causes that only a fractional amount of energy pass the coating. However, it is expectable that the EM waves, strongly attenuated, pass through regular metallic layers without special setups such as meshes or grids.

Introducing the *characteristic impedance* Z_0 which is defined as:

$$Z_0 = \frac{\hat{E}^+}{\hat{H}^+} = \hat{\eta} = \sqrt{\frac{j\omega\mu}{\sigma+j\omega\varepsilon}}$$ [2.2.20]

The impedance of vacuum is given by:

$$Z_{FS} = \sqrt{\frac{\mu_0}{\varepsilon_0}} = 120\pi\Omega \approx 377\Omega$$ [2.2.21]

Another important expression is the *loss tangent*, which can be written as:

$$\tan \delta = \frac{\sigma}{\omega \varepsilon} = \frac{\varepsilon''}{\varepsilon'}$$

[2.2.22]

The *loss tangent* is a common designation by manufacturers in specifying any material. The *loss tangent* can also be defined as the ratio of the imaginary and the real part of the complex permittivity. The complex permittivity can be written as:

$$\varepsilon' - j \cdot \varepsilon'' = \varepsilon' \cdot \left(1 - j \cdot \frac{\sigma}{\omega \varepsilon_0 \cdot \varepsilon'}\right)$$

[2.2.23]

With respect to the introduced equations in this section, the final time-domain forms of the electric and magnetic fields, for the pictured case in Figure 2-3, can be written as:

$$E_x = E^+ \cdot e^{-\alpha z} \cdot \cos(\omega t - \beta z + \theta^+) + E^- \cdot e^{\alpha z} \cdot \cos(\omega t + \beta z + \theta^-)$$

[2.2.24]

$$H_y = \frac{E^+}{\eta} \cdot e^{-\alpha z} \cdot \cos(\omega t - \beta z + \theta^+ - \theta_\eta) - \frac{E^-}{\eta} \cdot e^{\alpha z} \cdot \cos(\omega t + \beta z + \theta^- - \theta_\eta)$$

[2.2.25]

3. Propagation effects through different window constructions

This chapter presents several windowpane constructions, which are currently commonly used for domestic and public buildings. There exists a multitude of various window types for specific functions e.g. for safety, isolation purposes such as for heat, sound, fire and sun protection. To achieve these desired functions the manufacturers utilise special materials in the glass itself, extra glass layers and glass coatings. The investigations concerning wave propagation through windows are not very extensive up to now, because of the widespread assumption that the energy losses in a windowpane are negligibly small and that all windowpanes are identical in terms of propagation. That is true as long as the windowpanes only consist of glass. Modern windowpanes could have a bad influence on some EM waves, so that subscribers to some wireless services, such as mobile phone systems (GSM or UMTS) or satellite systems may not have sufficient signal strength inside buildings. The imminent arrival of new services, operating at higher frequencies, are dependent on an undisturbed penetration of the signal through windows, while a propagation through building walls, for instance brick walls, is not conceivable due to the high power losses [10].

The first part of this chapter deals with the history of glass, the most commonly used materials and the manufacturing processes. A theoretical model of the general propagation of EM waves through a single-pane and a multiple-pane construction is presented in section 3.2. Additionally, a brief description of the consequences for the wave propagation when coatings are components of a window is also included in this section. Chapter 3.3 describes the mathematical background of an EM wave passing through a multiple-pane window for vertical and horizontal polarisation. This section is the background for chapter 3.4, which refers to the software model. A summary about the ideas how the software works is also presented. The last section deals with conductive materials. Special attention has to be taken when those materials exist in windows. Therefore, some considerations are introduced which are necessary to obtain a correct model for wave propagation through them.

3.1. Modern window constructions

Glass is an amorphous and not a crystalline solid, which generally will be manufactured by melting processes. It is also known as a frozen, or undercooled fluid, because after the solidification of the melt there is not enough time left for crystal generation [11].

The glass industry has a history, which began in approximately 100 BC in the Roman Empire and also with the Egyptians. Some glass made artefacts have been discovered on the coasts of Ireland and Scotland, which were dated from the same time period. By coincidence, humans discovered the production of glass by using sand, the alkali from seaweed ash combined with high temperature. In the Middle East, the invention of glass was made by metal melting processes during the cooling of silica-rich slag.

The British scientist George Ravenscroft developed a glass with lead oxide in the seventeenth century. This glass has an individual refractive index, which encouraged other scientists to assume that various materials added to glass would cause other useful properties [12]. This was the beginning of the investigations to develop glass with various different characteristics.

After a lot of experiments and several glass compositions the current float glass, utilised for windowpane constructions, is the most commonly used glass type for construction purposes. It was invented by Alastair Pilkington in 1959, who also adopted a new method for manufacturing this specific glass type [13]. Float glass consists of modern materials. The largest part is still silica-oxide SiO_2 (over 70 %), Na_2O (sodium oxide with 14 %), CaO (calcium oxide with 9 %) and several other oxides such as aluminium and magnesium. For a variation of the refractive index the manufacturers still use lead and, additionally, baryta. Boron oxide is used to change the thermal and electrical properties. It is added to glass in a ratio of about 7% -15 %. The space between windowpanes is usually either evacuated or gas-filled (e.g. argon with the rest being air, krypton and xenon). Especially for windows with heat isolation characteristics, low emission coating layers, consisting of tin oxide, are used on the inside layer of the doubly glazed construction. Many different window construction types consist of layers or coatings to achieve anti-reflection effects, to absorb certain wavelengths (e.g. infrared, UV-light), or they are utilised for panes to achieve specific properties such as self-cleaning, safety and energy saving. These coatings are commonly made up of metal oxide, such as for sun protection [11]. But there are also layers consisting of ceramics. All of these layers can be coated on several sides of the window construction, thus the setup of a pane is variable.

The atomic design of glass is anisotropic, but this would have no big impact on the EM wave propagation, thus the influence of an irregular atomic setup is negligible for the frequency range of interest. These anisotropies emerge due to the tension distributions as a result of the fabrication process of heat-treated glasses. Criteria, which affect the assumption of glass being a homogeneous and isotropic media, are possibly observable irregularities in the windowpane. Therefore, a lot of requirements have to be performed, so that a pane is permitted to be used for construction purposes.

The German licence for windowpanes is allowed when factors like the following are fulfilled. Consider a windowpane with a surface area of about 1 metre squared, where a maximum of two inclusions, bubbles or spots are permitted with a diameter less than 2 mm. No marks longer than 15 mm on the pane surface are allowed and the waviness on a length of 300 mm does not exceed a value of 0.3 mm [14].

With respect to the requirements and the frequency range of the following investigations it is appropriate to regard a windowpane as smooth, isotropic, homogeneous and furthermore as an insulator due to the electrical resistivity- varying from 10^9 to 10^{20} $\Omega \cdot$cm. Therefore glass has a very low conductivity and finally good transmission characteristics, but only when no coatings are included.

3.2. Theoretical propagation model

The free-space wavelength of the frequencies concerning this thesis (for 90 MHz: 3.33 m and for 12.5 GHz: 2.4 cm) is large compared to the dimensions of the window construction. Despite the fact that the thickness of a windowpane is normally in the order of only a few millimetres, however, in addition to the harsh design criteria (see section 3.1), it is assumed the windowpane, consisting of dielectric materials, may be considered approximately homogeneous and isotropic. Therefore, the main propagation effects are reflection, refraction and transmission. This section shows, in a theoretical way, the anticipated complexity of the EM wave propagation through a single-pane and afterwards for a multiple-pane setup (with three panes). According to the current design of windowpanes, section 3.2.3 presents an assumption about the consequences when a windowpane consists of commonly utilised layers, coatings and gas filled space between the several panes.

3.2.1 Single - pane model

The single-pane model is suitable for an introduction into the transmission of EM waves through a window. The preview of this section already noted that the most frequently occurring effects will be reflection, refraction and transmission. Thus, Figure 3-1 explains the transmission paths of an incident wave more accurately, but consider that this is just an example for only one incident ray.

The red arrow points to the incident ray. When a uniform plane wave impinges on the surface of a windowpane, this scenario would be quite more complicated.

Power losses are not taken into account in this picture, but it is observable that all of the EM waves, which will be reflected back, carry a part of energy, which is definitely lost. Inside of the windowpane there are multiple-time reflections on both boundaries to the outdoor and indoor environments. The propagation effects are mathematically interpretable with the laws already described in chapter 2.

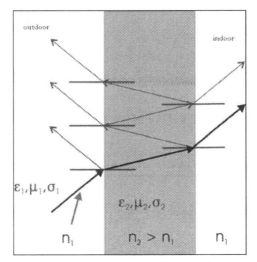

Fig. 3-1: Possible transmission path of an EM wave through a single-pane

3.2.2 Multiple - pane Model

A modern windowpane has multiple panes, due to the various possible functions they have to satisfy. Thus, if a comparison is made between the single-pane scenario and the triple-pane as shown in Figure 3-2, the transmission paths of EM waves can become complex.

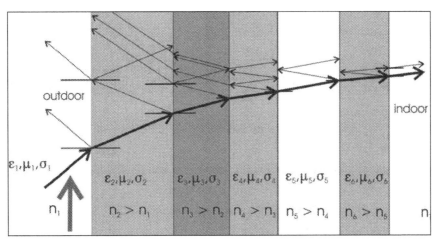

Fig. 3-2: Example for the wave propagation in a modern window pane

Similar to the previous chapter this is just an example and the red arrow is still pointing to the incident ray. There are many permutations such as multiple reflections in any layer, reflected waves from all layers back to the surface of the whole pane, and also these waves may be transmitted back into the window construction. The result, which is intimated in Figure 3-2, is a clutter of reflected rays, some of which will influence the received signal too, due to superposition. It is easy to note that due to this cluster of waves a considerable amount of energy could be lost. The magnitude of losses depends on the material properties of the window. It is important to note, however, that usually, for plain glass, the path losses are about 1 – 2 dB, so the propagating wave still accounts for approximately 65 % - 80 % of the original signal strength [15].

3.2.3 Window constructions with coatings and shielding

Many modern windowpanes also include shielding, dielectric layers and metallic coatings. These additional layers could disturb the transmission of EM waves through windows, due to the near-perfect reflection on some of their boundaries. Another possibility is when the window is designed with EM shielding, such as an electrically conductive mesh or metal foil connected to earth with a physical electrical connection (Faraday cage). This is sometimes desired, e.g. to maintain the data protection of a company so that nobody is able to get a prohibited connection to their wireless network. This could be realised for a whole frequency range [16]. Figure 3-3 illustrates how the setup of such a window looks like:

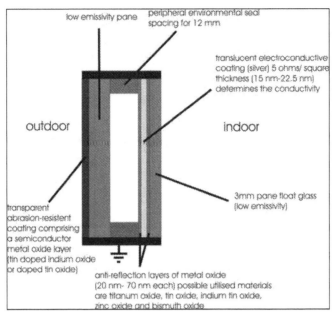

low emissivity pane peripheral environmental seal spacing for 12 mm

translucent electroconductive coating (silver) 5 ohms/ square thickness (15 nm-22.5 nm) determines the conductivity

outdoor   indoor

transparent abrasion-resistent coating comprising a semiconductor metal oxide layer (tin doped indium oxide or doped tin oxide)

3mm pane float glass (low emissivity)

anti-reflection layers of metal oxide (20 nm- 70 nm each) possible utilised materials are titanium oxide, tin oxide, indium tin oxide, zinc oxide and bismuth oxide

Fig. 3-3: Example for a window with electromagnetic shielding [16]

A US Patent [17] presents an interesting idea of a window design, which can absorb microwave energy at selective frequencies. This window, which is installed in vehicles for instance in airplanes, consists of two dielectric layers, with electrically conductive coatings on each dielectric layer surface. Behind that window is a sensor which is utilised for other purposes. But this sensor has to be protected against certain radiation to prevent a restriction in functionality, such as spurious readings. The sensor also could be damaged by high energy microwaves. Therefore, the window is responsible for absorbing particular frequencies so that no reflection of microwaves can occur. In addition, these microwaves cannot pass the window and reach the sensor. The idea is that for instance a military aircraft cannot be detected by radar due to the reflectivity of windows. Figure 3-4 presents the setup of the coatings and how this window is constructed.

The grid size of the conductive layers is less or much less than the wavelength of the microwave energy that is to be reflected. The average dimensions of the shape sizes are between 5 and 15 micrometers. The spacing between the shapes varies between 0.1 and 2 millimetres. Signals in the frequency ranges of the L-Band, S-Band, C-Band, X-Band and the K_a–Band can be effectively influenced by this construction. Figure 3-5 illustrates the attenuations of reflected and transmitted waves as a function of the microwave frequency.

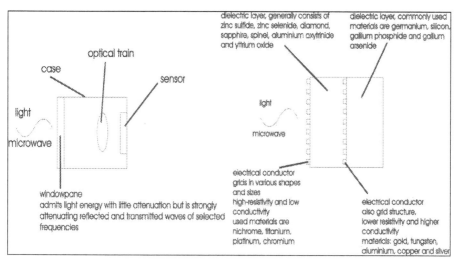

Fig. 3-4:   Setup of frequency absorbing window constructions [17]

In another case, it is possible to develop coatings, which are conceived for a good transmission of certain frequencies, such as for GSM [18]. Similar to the patent [17], the investigation of the University of Technology in Lund, Sweden [18], is also based on a metal grid in the glass with a constant spacing and various shapes. Depending on the grid size, shapes, thickness and the materials it is possible to let the waves of the frequency range of interest pass through the glass, be absorbed or be reflected by it. Usual shapes are for instance circular, rectangular dipoles and crossed dipoles.

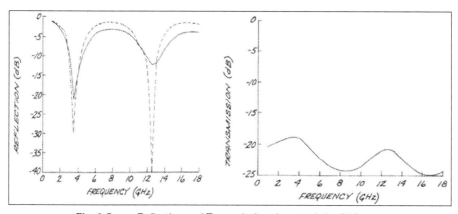

Fig. 3-5:   Reflection and Transmission characteristics [17]

These specific layers are named as *Frequency Selective Surfaces* (FSS), and operate in the same way as microwave filters for millimetre waves. They may have low-pass and high-pass characteristics. The principle of a FSS is that the dipole elements will resonate when an EM wave impinges on the surface. For that, the half-wavelength of the signal has to be in the order of a multiple times of the length of the dipoles, because the dipoles are comparable with half-wavelength dipole antennas. Due to the EM wave induced surface current, each strip of the layer has a phase delay. The result is that the overall reradiated energy will propagate in the same angle away from the FSS as the reflected wave would propagate. When the wavelength of the signal deviates from the shape size or the dimensions of the dipoles, the EM waves pass the FSS, affected by small losses only. Figure 3-6 presents an example for one of these layers, while Figure 3-7 shows two graphs where the necessary sizes of the element lengths over the frequency is presented and, additionally, an example for a FSS, which does not let EM waves pass through it with a frequency of 1.9 GHz [19].

A practical example is the US – company GlassLock. They developed window coatings called SpyGuard[TM], which reflect standard frequencies such as WLAN. For preventing undesired access to wireless networks by others, these coatings provide a high attenuation on signals, propagating through a window from both sides [20].

Finally, it can be said that multiple technical applications may be considered through selective additions such as coatings. They can, however, also cause other undesirable effects, which can be solved exactly with the same tools, coatings and shielding. The investigations in terms of wave transmission make researchers not only go after new technical and electrical problems, but they additionally try to solve other problems, such as for data protection and for military applications.

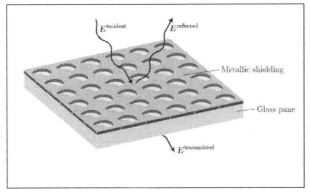

Fig. 3-6:   Frequency Selective Surface [18]

In terms of this thesis, an investigation of wave transmission in the manner outlined in sections 3.2.1 and 3.2.2, with panes consisting of additions, as also presented briefly in the reports [16, 17, 18] already mentioned, would cause very complex effects, which cannot be investigated with the assumptions described earlier. Thus, the analysis will be made with regular panes, where every pane will have its individual material characteristics. Therefore, section 3.5 finally presents in a theoretical way what consequences the conductivity of materials has on the reflectivity and transmission of EM waves for different frequencies. It is predictable that the extra coatings would cause the highest energy losses in the window construction due to reflection and absorption. The dependence on the frequency, and finally, estimations about how accurate the assumptions for the math model are, is very important to investigate.

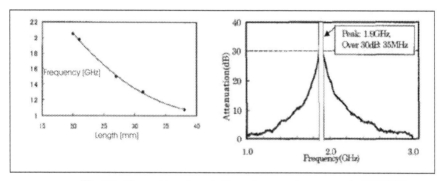

Fig. 3-7: Element length and attenuation over frequency of a FSS [19]

3.3. General approach for predicting wave propagation

A wave originating in air (lossless medium) is incident with an oblique angle upon a planar surface of a windowpane, which is shown in Figure 3-8 and Figure 3-9. Referring to section 3.1, note that the windowpane is assumed to be an insulator due to the very low conductivity of glass. In fact, glass is defined as a dielectric material. According to Balanis [3], the conductivity is in the order of 10^{-12} S/m and the relative permeability is approximately unity. Hence, the *attenuation factor* α is negligibly small.

Nonetheless, the following equations include these parameters due to the frequency dependence of the permittivity, and also as modern window constructions have layers, which can contain materials that contribute to power

loss. The theoretical presentation of the math model in section 3.4 supports reflection and refraction effects for regular materials only. Lossy materials such as windowpanes with gas-filled spacing or with irregular coatings presented in [18] are in terms of the measurement series, presented in chapter 4 and 5, not considered, due to the actual windows being analysed. Despite the investigations with a heat insulating window, a simulation cannot be done due to the unknown gas utilised and pane coatings.

This section introduces linearly polarisations and demonstrates for both types the mathematical description of the EM wave propagation through a window construction.

3.3.1 Horizontal polarisation

The definition of the polarisation type depends on the E-field orientation. For the case of the horizontal (or perpendicular) polarisation the E-field vector is perpendicular to the *plane of incidence* which Figure 3-8 shows.

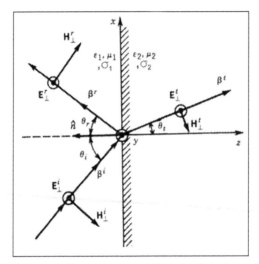

Fig. 3-8:   Horizontal (or perpendicular) polarisation [3]

It is appropriate to adopt new parameters, which are called the *reflection coefficient, Γ*, and *transmission coefficient, T*. These coefficients describe the reflection and transmission of the EM wave's fields at a boundary between two media. They depend on the angle of incidence, both material properties and the polarisation. For horizontal polarisation these coefficients are defined as:

$$\Gamma_{\perp} = \frac{E_r}{E_i} = \frac{\eta_2 \cdot \cos\theta_i - \eta_1 \cdot \cos\theta_t}{\eta_2 \cdot \cos\theta_i + \eta_1 \cdot \cos\theta_t} \qquad\qquad [3.3.1]$$

The *transmission coefficient* can be obtained from:

$$T_\perp = \frac{E_t}{E_i} = \frac{2 \cdot \eta_2 \cdot \cos \theta_i}{\eta_2 \cdot \cos \theta_i + \eta_1 \cdot \cos \theta_t} \qquad [3.3.2]$$

As shown in Figure 3-8 the EM wave moves in the x-z-plane. To analyse the incident, reflected and transmitted angles, the electromagnetic fields as well as the propagation constant have to be decomposed into the perpendicular (x) and parallel (z) components. The total reflected and transmitted field can be obtained by calculating the vector sum from each of these two components [3].

The E-field vector is perpendicular to the *plane of incidence*. The incident and reflected fields are just travelling in free space, so the *attenuation factor* α can be set to zero for medium 1. To calculate these components we can write:

$$E_\perp^i = \hat{a}_y \cdot E_0 \cdot e^{-j\beta_1(x \cdot \sin \theta_i + z \cdot \cos \theta_i)} \qquad [3.3.3]$$

$$H_\perp^i = (-\hat{a}_x \cdot \cos \theta_i + \hat{a}_z \cdot \sin \theta_i) \cdot \frac{E_0}{\eta_1} \cdot e^{-j\beta_1(x \cdot \sin \theta_i + z \cdot \cos \theta_i)} \qquad [3.3.4]$$

Relating to Figure 3-8, $\hat{a}_n$ represents the directions of the field vectors, respectively. The phase constant β describes the direction of propagation in terms of the incident angle θ_i, decomposed in two components. The magnetic field vector is appropriately expressed in terms of the ratio of the amplitude of the electric field and the characteristic impedance of the relevant medium.

The reflected components are defined as:

$$E_\perp^r = \hat{a}_y \cdot \Gamma_\perp \cdot E_0 \cdot e^{-j\beta_1(x \cdot \sin \theta_r - z \cdot \cos \theta_r)} \qquad [3.3.5]$$

$$H_\perp^r = (\hat{a}_x \cdot \cos \theta_r + \hat{a}_z \cdot \sin \theta_r) \cdot \frac{\Gamma_\perp \cdot E_0}{\eta_1} \cdot e^{-j\beta_1(x \cdot \sin \theta_r - z \cdot \cos \theta_r)} \qquad [3.3.6]$$

Just as for the incident fields, in Figure 3-8, the reflected field components, propagating, of course, in the opposite direction to the incident field components, have to be modified in that the sign of all z-components needs to change. Additionally, the amplitudes of the field vectors are modified by the reflection coefficient, obtainable with equation 3.3.1. The reflected angle θ_r is equal to the incident angle θ_i.

Finally the electric and the magnetic field of the transmitted wave, including the *attenuation factor* α of the particular medium may be written as:

$$E'_\perp = \hat{a}_y \cdot T_\perp \cdot E_0 \cdot e^{\alpha_2(x \cdot \sin\theta_t + z \cdot \cos\theta_t)} \cdot e^{-j\beta_2(x \cdot \sin\theta_t + z \cdot \cos\theta_t)} \qquad [3.3.7]$$

$$H'_\perp = (-\hat{a}_x \cdot \cos\theta_t + \hat{a}_z \cdot \sin\theta_t) \cdot \frac{T_\perp \cdot E_0}{\eta_2} \cdot e^{-\alpha_2(x \cdot \sin\theta_t + z \cdot \cos\theta_t)} \cdot e^{-j\beta_2(x \cdot \sin\theta_t + z \cdot \cos\theta_t)} \qquad [3.3.8]$$

The parameters of the other medium have to be utilised now. The transmission coefficient, which can be calculated with equation 3.3.2, and the refraction angle must be included in these equations. The amplitude of bold fields decreases with penetration into the medium. Thus, the sign of the attenuation coefficient α is negative.

We considered the windowpane to be a dielectric material. Balanis [3] simplified the general equations 2.2.17 and 2.2.18 for the *attenuation factor,* α, and the *phase constant* β for the case of a dielectric material. The stipulation for a dielectric, which is equal to the imaginary part of the complex permittivity already presented with equation 2.2.23, is defined as:

$$\frac{\sigma}{\omega\varepsilon} \ll 1 \qquad [3.3.9]$$

Baker [21] investigated different measurement techniques, which are applicable for the characterisation of low loss materials. During the comparison of several measurement procedures, he received parameters for different glass types such as the real part of the complex permittivity and the loss tangent. The imaginary parts of the different glass types are not distinguishable in its tiny value, which is in the order of about 0.0002. The loss tangent is around 6.8×10^{-5}, which defines glass as a low loss material. Baker defines a material as being lossy, when its loss tangent is in the order of 0.005 [21]. The loss tangent will increase above several GHz, but its value is too tiny, therefore it has not to be taken into account at these frequencies as well.

The *attenuation factor* α for this special case is equal to:

$$\alpha = \frac{\sigma}{2} \cdot \sqrt{\frac{\mu}{\varepsilon}}$$

[3.3.10]

Due to the very low conductivity it is apparent that α is very close to zero. Note that the permeability of glass is assumed to be equal to the permeability of free space, μ_0. Nonetheless the parameter μ_r and α will be kept in the following equations due to the potential inclusion of metallic additions in the window construction and the variation of the permittivity for different frequencies. Accordingly the *phase constant* β can be represented by a simplified equation too:

$$\beta = \omega \cdot \sqrt{\mu \cdot \varepsilon}$$

[3.3.11]

The *reflection coefficient* $\Gamma_\perp$ and the *transmission coefficient* $T_\perp$ can be obtained from equations 3.3.1 and 3.3.2. The simplifications of the *characteristic impedances* can be written as:

$$\eta_1 = \sqrt{\frac{\mu_0}{\varepsilon_0}}$$

[3.3.12]

$$\eta_2 = \sqrt{\frac{\mu_0 \cdot \mu_r}{\varepsilon_0 \cdot \varepsilon_r}}$$

[3.3.13]

The general equation for the *characteristic impedance* is equation 2.2.20. Index 1 represents the free space medium and index 2 designates the impedance of the windowpane. Table 3-1 illustrates an overview of the complete equations to determine the *propagation constant* γ, the *intrinsic impedance* η, the *wavelength* λ, the *phase velocity* v_P and *skin depth* δ for all lossy materials [3].

The conditions for determining whether the material is a dielectric or a conductor can be determined by *Ampère's law* (equation 2.2.2) for materials, which could have current and charge densities. *Ampère's law* can be written as:

$$\nabla \times H = J_c + J_d = \sigma E + j\omega\varepsilon E = (\sigma + j\omega\varepsilon)E$$

[3.3.14]

J_c represents the conduction, and J_d the displacement current densities. For a good dielectric, the displacement current density is much higher in value than the conduction current density. This is the case, when equation 3.3.9 is valid. If we are dealing with a good conductor, the conduction current density is, of course, greater than the displacement current density.

The determination, whether the material is a dielectric or a conductor, differs in several references. The following determination is also used, where the material properties' expression $\left(\dfrac{\sigma}{\omega\varepsilon}\right)^{2} \ll 1$, i.e. the result must be very small compared to unity. Therefore, it is important to note that the values of these arguments have to be extremely small and should be compared with the equations for exact solutions too, to ensure that no additional inaccuracy will occur, due to math simplifications.

Making use of the boundary conditions of two media we can get a relationship between both coefficients. Since for a source and charge free boundary between two media, the tangential components of the electromagnetic fields are continuous [3], it is legitimate to derive following equations at the interface:

$$(E_\perp^i + E_\perp^r)\,|_{z=0} = E_\perp^t\,|_{z=0} \qquad\qquad [3.3.15]$$

$$(H_\perp^i + H_\perp^r)\,|_{z=0} = H_\perp^t\,|_{z=0} \qquad\qquad [3.3.16]$$

	Exact	Good dielectric $\dfrac{\sigma}{\omega\varepsilon}\ll 1$	Good conductor $\dfrac{\sigma}{\omega\varepsilon}\gg 1$
attenuation constant α	$=\omega\sqrt{\mu\varepsilon}\sqrt{\dfrac{1}{2}\left[\sqrt{1+\left(\dfrac{\sigma}{\omega\varepsilon}\right)^2}-1\right]}$	$\approx\dfrac{\sigma}{2}\cdot\sqrt{\dfrac{\mu}{\varepsilon}}$	$\approx\sqrt{\dfrac{\omega\mu\sigma}{2}}$
phase constant β	$=\omega\sqrt{\mu\varepsilon}\sqrt{\dfrac{1}{2}\left[\sqrt{1+\left(\dfrac{\sigma}{\omega\varepsilon}\right)^2}+1\right]}$	$\approx\omega\sqrt{\mu\varepsilon}$	$\approx\sqrt{\dfrac{\omega\mu\sigma}{2}}$
characteristic impedance η	$=\sqrt{\dfrac{j\omega\mu}{\sigma+j\omega\varepsilon}}$	$\approx\sqrt{\dfrac{\mu}{\varepsilon}}$	$\approx\sqrt{\dfrac{\omega\mu}{2\sigma}}(1+j)$
wavelength λ	$=\dfrac{2\pi}{\beta}$	$\approx\dfrac{2\pi}{\omega\sqrt{\mu\varepsilon}}$	$\approx 2\pi\sqrt{\dfrac{2}{\omega\mu\sigma}}$
phase velocity v	$=\dfrac{\omega}{\beta}$	$\approx\dfrac{1}{\sqrt{\mu\varepsilon}}$	$\approx\sqrt{\dfrac{2\omega}{\mu\sigma}}$
skin depth δ	$=\dfrac{1}{\alpha}$	$\approx\dfrac{2}{\sigma}\cdot\sqrt{\dfrac{\varepsilon}{\mu}}$	$\approx\sqrt{\dfrac{2}{\omega\mu\sigma}}$

Table 3-1:   TEM wave in lossy media [3]

From equations 3.3.3, 3.3.5 and 3.3.15 we obtain (note that z=0):

$$\hat{a}_y\cdot E_0\cdot e^{-j\beta_1(x\cdot\sin\theta_i)}+\hat{a}_y\cdot\Gamma_\perp\cdot E_0\cdot e^{-j\beta_1(x\cdot\sin\theta_r)}$$
$$=\hat{a}_y\cdot T_\perp\cdot E_0\cdot e^{-\alpha_2(x\cdot\sin\theta_t)}\cdot e^{-j\beta_2(x\cdot\sin\theta_t)}$$

[3.3.17]

and equations 3.3.4, 3.3.6 and 3.3.16 delivers:

$$(-\hat{a}_x\cdot\cos\theta_i)\cdot\frac{E_0}{\eta_1}\cdot e^{-j\beta_1(x\cdot\sin\theta_i)}+(\hat{a}_x\cdot\cos\theta_r)\cdot\frac{\Gamma_\perp\cdot E_0}{\eta_1}\cdot e^{-j\beta_1(x\cdot\sin\theta_r)}$$
$$=(-\hat{a}_x\cdot\cos\theta_t)\cdot\frac{T_\perp\cdot E_0}{\eta_2}\cdot e^{-\alpha_2(x\cdot\sin\theta_t)}\cdot e^{-j\beta_2(x\cdot\sin\theta_t)}$$

[3.3.18]

These equations are divided into their real and imaginary parts. It is only necessary to look after the real part of the solution, because the imaginary part represents the reactive power of the EM wave, which has no influence on the amount of the transmitted power. After simplifying both equations, we can write for equation 3.3.17:

$$\cos(\beta_1 \cdot x \cdot \sin\theta_i) + \Gamma_\perp \cdot \cos(\beta_1 \cdot x \cdot \sin\theta_r) = T_\perp \cdot \cos((-\alpha_2 + \beta_2) \cdot x \cdot \sin\theta_t) \qquad [3.3.19]$$

Similar to the expression for the electric field, equation 3.3.19, we obtain for the real part of the magnetic field:

$$\frac{1}{\eta_1} \cdot \left[-\cos\theta_i \cdot \cos(\beta_1 \cdot x \cdot \sin\theta_i) + \Gamma_\perp \cdot \cos\theta_r \cdot \cos(\beta_1 \cdot x \cdot \sin\theta_r) \right]$$
$$= -\frac{1}{\eta_2} \cdot T_\perp \cdot \cos\theta_t \cdot \cos((-\alpha_2 + \beta_2) \cdot x \cdot \sin\theta_t) \qquad [3.3.20]$$

With *Snell's Law of Reflection* and *Law of Refraction* already presented, the following relations can be used in equations 3.3.19 and 3.3.20:

$$\theta_i = \theta_r \qquad [3.3.21]$$

$$\beta_1 \cdot \sin\theta_i = \beta_2 \cdot \sin\theta_t \qquad [3.3.22]$$

To achieve a convenient overview about the relationship between the *reflection coefficient* and the *transmission coefficient* we considered a simple windowpane without any additions up to now. At this point, it is important to weigh how many losses signals have when propagating through a window. Because information about the permittivity of glass for a certain frequency range is in most of the cases not available, the attenuation has primarily to be obtained by practical investigations. Therefore, for simplification purposes, the attenuation factor is set to zero. But this approach has to be modified when the analysis of the wave propagation will obviously show different results. Referring to equation 3.3.11, the equation 3.3.19 becomes:

$$1 + \Gamma_\perp = T_\perp \qquad [3.3.23]$$

and equation 3.3.20 can be written as:

$$\frac{\cos\theta_i}{\eta_1}\cdot(\Gamma_\perp-1)=-\frac{T_\perp}{\eta_2}\cdot\cos\theta_t \qquad [3.3.24]$$

With these equations it is easy to derive the equations 3.3.1 and 3.3.2. The equations for the *reflection coefficient* $\Gamma_\perp$ and the *transmission coefficient* $T_\perp$ can also be rewritten for the case that the permeability of both media is equal. To obtain equations, which only depend on the angle of incidence and the material properties of both media, it is necessary to consider also equations 3.3.12 and 3.3.13. Additionally, the following relation is needed:

$$\sin^2\theta+\cos^2\theta=1 \qquad [3.3.25]$$

Finally, the equations for $\Gamma_\perp$ and $T_\perp$ are:

$$\Gamma_\perp=\frac{\cos\theta_i-\sqrt{\dfrac{\varepsilon_2}{\varepsilon_1}}\cdot\sqrt{1-\left(\dfrac{\varepsilon_1}{\varepsilon_2}\right)\cdot\sin^2\theta_i}}{\cos\theta_i+\sqrt{\dfrac{\varepsilon_2}{\varepsilon_1}}\cdot\sqrt{1-\left(\dfrac{\varepsilon_1}{\varepsilon_2}\right)\cdot\sin^2\theta_i}} \qquad [3.3.26]$$

$$T_\perp=\frac{2\cdot\cos\theta_i}{\cos\theta_i+\sqrt{\dfrac{\varepsilon_2}{\varepsilon_1}}\cdot\sqrt{1-\left(\dfrac{\varepsilon_1}{\varepsilon_2}\right)\cdot\sin^2\theta_i}} \qquad [3.3.27]$$

The transmitted power of a horizontal polarised EM wave in the far-field region, which penetrates into the simple windowpane without additions, is determined with the equations 3.3.7, 3.3.8 and 2.2.14:

$$S_{\perp av}^{t} = \frac{1}{2} \cdot Re \cdot (E_{\perp}^{t} \times H_{\perp}^{t*})$$

$$= \frac{1}{2} \cdot (\hat{a}_{y} \cdot T_{\perp} \cdot E_{0} \cdot e^{-j\beta_{2}(x \cdot \sin\theta_{t} + z \cdot \cos\theta_{t})} \times$$

$$(-\hat{a}_{x} \cdot \cos\theta_{t} + \hat{a}_{z} \cdot \sin\theta_{t}) \cdot \frac{T_{\perp} \cdot E_{0}}{\eta_{2}} \cdot e^{+j\beta_{2}(x \cdot \sin\theta_{t} + z \cdot \cos\theta_{t})}) \qquad [3.3.28]$$

$$= \hat{a}_{x} \cdot \left(\sin\theta_{t} \cdot \frac{|T_{\perp}|^{2} \cdot |E_{0}|^{2}}{2 \cdot \eta_{2}} \right) + \hat{a}_{z} \cdot \left(\cos\theta_{t} \cdot \frac{|T_{\perp}|^{2} \cdot |E_{0}|^{2}}{2 \cdot \eta_{2}} \right)$$

$$= \frac{|T_{\perp}|^{2} \cdot |E_{0}|^{2}}{2 \cdot \eta_{2}} \cdot (\hat{a}_{x} \cdot \sin\theta_{t} + \hat{a}_{z} \cdot \cos\theta_{t})$$

Also, it is appropriate to show that, in general, for most of all dielectric materials such as glass no *Brewster Angle* ($\Gamma = 0$) for horizontal polarisation exists. The equation for the angle of total transmission is given by:

$$\sin\theta_{i} = \sqrt{\frac{\dfrac{\varepsilon_{2}}{\varepsilon_{1}} - \dfrac{\mu_{2}}{\mu_{1}}}{\dfrac{\mu_{1}}{\mu_{2}} - \dfrac{\mu_{2}}{\mu_{1}}}} \qquad [3.3.29]$$

The sine function cannot be unity. Therefore if μ_{1} is equal to μ_{2} then the denominator is zero and the value of the sine function is infinity. The conclusion is that the *reflection coefficient* will never be zero. Hence, no total transmission in the case of horizontal polarisation is anticipated. Of course, if the permeabilities of both media are different, total transmission is possible for a certain incident angle.

Total reflection occurs when the following relation is satisfied:

$$\sin\theta_{c} \geq \sqrt{\frac{\mu_{2}\varepsilon_{2}}{\mu_{1}\varepsilon_{1}}} \qquad [3.3.30]$$

This is only possible when the wave propagates from a more dense to a less dense medium, which means that $\varepsilon_{2} < \varepsilon_{1}$. For instance, the wave propagates from glass to air. Note that total reflection for angles above θ_{c} is also occurring. For the case of total reflection, the transmitted wave will propagate along the surface of the boundary, so that the EM waves are perpendicular to the surface normal. Hence, this wave is called surface wave, which will not transport any energy in the z-direction.

3.3.2 Vertical polarisation

The E-field vector lies in the plane of incidence, which is called vertical (or parallel) polarisation. Figure 3-9 illustrates that case:

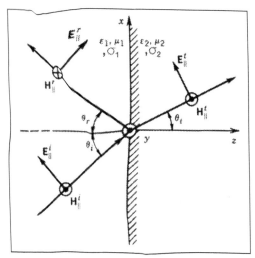

Fig. 3-9:   Vertical (or parallel) polarisation [3]

The *reflection coefficient* is given by:

$$\Gamma_{\parallel} = \frac{E_r}{E_i} = \frac{\eta_2 \cdot \cos\theta_t - \eta_1 \cdot \cos\theta_i}{\eta_2 \cdot \cos\theta_t + \eta_1 \cdot \cos\theta_i} \qquad [3.3.31]$$

and the *transmission coefficient* can be written as:

$$T_{\parallel} = \frac{E_t}{E_i} = \frac{2 \cdot \eta_2 \cdot \cos\theta_i}{\eta_1 \cdot \cos\theta_i + \eta_2 \cdot \cos\theta_t} \qquad [3.3.32]$$

Similar to section 3.3.1, this section will describe the electromagnetic fields with the assumptions already made for this kind of polarisation. With respect to Figure 3-9 we can write for the field components:

$$E_{\parallel}^i = (\hat{a}_x \cdot \cos\theta_i - \hat{a}_z \cdot \sin\theta_i) \cdot E_0 \cdot e^{-j\beta_1(x\cdot\sin\theta_i + z\cdot\cos\theta_i)}$$

[3.3.33]

$$H_{\parallel}^i = \hat{a}_y \cdot \frac{E_0}{\eta_1} \cdot e^{-j\beta_1(x\cdot\sin\theta_i + z\cdot\cos\theta_i)}$$

[3.3.34]

The reflected field components can be obtained from:

$$E_{\parallel}^r = (\hat{a}_x \cdot \cos\theta_r + \hat{a}_z \cdot \sin\theta_r) \cdot \Gamma_{\parallel} \cdot E_0 \cdot e^{-j\beta_1(x\cdot\sin\theta_r - z\cdot\cos\theta_r)}$$

[3.3.35]

$$H_{\parallel}^r = -\hat{a}_y \cdot \frac{\Gamma_{\parallel} \cdot E_0}{\eta_1} \cdot e^{-j\beta_1(x\cdot\sin\theta_r - z\cdot\cos\theta_r)}$$

[3.3.36]

And finally the transmitted components of both electromagnetic fields:

$$E_{\parallel}^t = (\hat{a}_x \cdot \cos\theta_t - \hat{a}_z \cdot \sin\theta_t) \cdot T_{\parallel} \cdot E_0 \cdot e^{-\alpha_2(x\cdot\sin\theta_t + z\cdot\cos\theta_t)} \cdot e^{-j\beta_2(x\cdot\sin\theta_t + z\cdot\cos\theta_t)}$$

[3.3.37]

$$H_{\parallel}^t = \hat{a}_y \cdot \frac{T_{\parallel} \cdot E_0}{\eta_2} \cdot e^{-\alpha_2(x\cdot\sin\theta_t + z\cdot\cos\theta_t)} \cdot e^{-j\beta_2(x\cdot\sin\theta_t + z\cdot\cos\theta_t)}$$

[3.3.38]

With the same simplifications made in equations from 3.3.9 to 3.3.16 we derive the equations for both electromagnetic fields for the boundary conditions at z=0:

$$(\hat{a}_x \cdot \cos\theta_i) \cdot E_0 \cdot e^{-j\beta_1(x\cdot\sin\theta_i)} + (\hat{a}_x \cdot \cos\theta_r) \cdot \Gamma_{\parallel} \cdot E_0 \cdot e^{-j\beta_1(x\cdot\sin\theta_r)}$$
$$= (\hat{a}_x \cdot \cos\theta_t) \cdot T_{\parallel} \cdot E_0 \cdot e^{-\alpha_2(x\cdot\sin\theta_t)} \cdot e^{-j\beta_2(x\cdot\sin\theta_t)}$$

[3.3.39]

$$\hat{a}_y \cdot \frac{E_0}{\eta_1} \cdot e^{-j\beta_1(x\cdot\sin\theta_i)} - \hat{a}_y \cdot \frac{\Gamma_{\parallel} \cdot E_0}{\eta_1} \cdot e^{-j\beta_1(x\cdot\sin\theta_r)}$$
$$= \hat{a}_y \cdot \frac{T_{\parallel} \cdot E_0}{\eta_2} \cdot e^{-\alpha_2(x\cdot\sin\theta_t)} \cdot e^{-j\beta_2(x\cdot\sin\theta_t)}$$

[3.3.40]

After the separation of the real parts of equations 3.3.39 and 3.3.40 we can write:

$$\cos\theta_i \cdot \cos(\beta_1 \cdot x \cdot \sin\theta_i) + \cos\theta_r \cdot \Gamma_{\parallel} \cdot \cos(\beta_1 \cdot x \cdot \sin\theta_r)$$
$$= \cos\theta_t \cdot T_{\parallel} \cdot \cos((-\alpha_2 + \beta_2) \cdot x \cdot \sin\theta_t)$$

[3.3.41]

$$\frac{1}{\eta_1} \cdot \left[\cos(\beta_1 \cdot x \cdot \sin\theta_i) - \Gamma_{\parallel} \cdot \cos(\beta_1 \cdot x \cdot \sin\theta_r)\right]$$
$$= \frac{1}{\eta_2} \cdot T_{\parallel} \cdot \cos((-\alpha_2 + \beta_2) \cdot x \cdot \sin\theta_t)$$

[3.3.42]

Referring to equations 3.3.21 and 3.3.22 we obtain the following relation between the *reflection coefficient* and the *transmission coefficient*:

$$\frac{\cos\theta_i}{\cos\theta_t} \cdot \left(1 + \Gamma_{\parallel}\right) = T_{\parallel}$$

[3.3.43]

$$\frac{\eta_2}{\eta_1}(1 - \Gamma_{\parallel}) = T_{\parallel}$$

[3.3.44]

Similar to section 3.3.1 the expressions in 3.3.43 and 3.3.44 will be rewritten for the case that the permeabilities are equal in both media. With respect to equations 3.3.31 and 3.3.32 we derive:

$$\Gamma_{\parallel} = \frac{-\cos\theta_i + \sqrt{\frac{\varepsilon_1}{\varepsilon_2}} \cdot \sqrt{1 - \left(\frac{\varepsilon_1}{\varepsilon_2}\right) \cdot \sin^2\theta_i}}{\cos\theta_i + \sqrt{\frac{\varepsilon_1}{\varepsilon_2}} \cdot \sqrt{1 - \left(\frac{\varepsilon_1}{\varepsilon_2}\right) \cdot \sin^2\theta_i}}$$

[3.3.45]

$$T_{\parallel} = \frac{2 \cdot \sqrt{\frac{\varepsilon_1}{\varepsilon_2}} \cdot \cos\theta_i}{\cos\theta_i + \sqrt{\frac{\varepsilon_1}{\varepsilon_2}} \cdot \sqrt{1 - \left(\frac{\varepsilon_1}{\varepsilon_2}\right) \cdot \sin^2\theta_i}}$$

[3.3.46]

The transmitted power of a vertically polarised EM wave in the far-field region can be obtained with equations 3.3.37, 3.3.38 (where α is set to zero) and 2.2.14:

$$S^t_{\|av} = \frac{1}{2} \cdot Re(E^t_{\perp} \times H^{t*}_{\perp})$$

$$= \frac{1}{2} \cdot ((\hat{a}_x \cdot \cos\theta_t - \hat{a}_z \cdot \sin\theta_t) \cdot T_{\|} \cdot E_0 \cdot e^{-j\beta_2(x \cdot \sin\theta_t + z \cdot \cos\theta_t)} \times$$

$$\hat{a}_y \cdot \frac{T_{\|} \cdot E_0}{\eta_2} \cdot e^{+j\beta_2(x \cdot \sin\theta_t + z \cdot \cos\theta_t)}) \hspace{3cm} [3.3.47]$$

$$= \hat{a}_x \cdot \left(\frac{|T_{\|}|^2 \cdot |E_0|^2}{2 \cdot \eta_2} \cdot \sin\theta_t \right) + \hat{a}_z \cdot \left(\frac{|T_{\|}|^2 \cdot |E_0|^2}{2 \cdot \eta_2} \cdot \cos\theta_t \right)$$

$$= \frac{|T_{\|}|^2 \cdot |E_0|^2}{2 \cdot \eta_2} (\hat{a}_z \cdot \cos\theta_t + \hat{a}_x \cdot \sin\theta_t)$$

Unlike horizontal polarisation, a *Brewster Angle* exists for vertical polarised EM waves. The *Brewster Angle* is, with respect to the same assumption made for equation 3.3.29, shown in the following expression:

$$\sin\theta_i = \sin\theta_h = \sqrt{\frac{\varepsilon_1}{\varepsilon_1 + \varepsilon_2}} \hspace{3cm} (3.3.44)$$

Finally, the condition for total reflection of vertically polarised EM waves is the same as for horizontally polarised waves. Thus, equation 3.3.30 is also valid, taking into account all declared limitations, already described for horizontal polarisation.

3.4. Completing the math model: parameters and assumptions

For the confirmation of the achieved measurement results, a software model is convenient, which supports the propagation effects described in chapter 2.1. The development of the software bases on the principle that the signal losses of a single EM wave will be calculated when it passes the window construction. Despite that only the path losses of approximately normally incident EM wavefronts through the specimen are practically measured, the software additionally computes different incident angles too. A simulation of whole uniform plane waves was considered first, but the computation time was very high and the reliability was not sure, but a comparison of the simulation and the measurement results showed that it is sufficient to deal with just one impinging EM wave. So, afterwards, simplifications in the source code have been made. This chapter presents the algorithm of the model.

3.4.1 Input parameters

The developed model only considers dielectric materials. Therefore, the input parameters are just the incident angle, the type of polarisation and the permittivity of the layers of the window construction. With respect to Table 3-1 and the column for materials with good dielectric properties, it is apparent that it is not necessary to consider the frequency of the wave as an input parameter. But that is only valid for window constructions without metallic coatings. However, the only parameter, which can be described by the wavelength, is the *phase constant β*. This parameter is only necessary when calculating the refracted angle of an EM wave. But the wavelength will be cancelled during the calculation. Only the *reflection*, the *transmission coefficient* and the incident angle are responsible for the losses the EM wave will be affected while propagating through the layers (Equations 3.3.26, 3.3.27, 3.3.45, and 3.3.46).

3.4.2 Algorithm

For all possible boundaries, the *transmission* and *reflection coefficient* in terms of the polarisation will be calculated and the power density as well. The last mentioned delivers the transmitted energy through the window. The applied equations are already mentioned in section 3.3.

3.5. Conductive materials

Conductivity is responsible for the losses in materials, because time-varying fields induce currents. As shown in chapter 3.2.3, many windows consist of very thin layers made of conductive, semiconductive and sometimes ceramic materials. In contrast, glass is assumed to be an insulator because of the negligibly small conductivity and high resistance. Despite the fact that the wave propagation through glass is from the mathematical point of view lossless, because the *attenuation factor α* is very close to zero, many measurements have, however, demonstrated different, if not opposite, results. The losses in a windowpane are namely very small. But including the losses of the conductive layers, the overall losses of a window construction can be considerable. The influence of the good reflectivity of these layers, which is already presented in chapter 3.2.3, can be quite high, due to the development of window constructions, which try to deliberately exploit this aspect. As the frequency of the EM waves varies, it is not only the permittivity of glass that may change. The *attenuation factor, α*, may also increase and thus, it would be expected that the losses would get higher with increasing frequency. Figure 3-10 illustrates the theoretical changing of the *attenuation factor*, α, over the frequency, calculated from Balanis [3] for the static conductivity of iron.

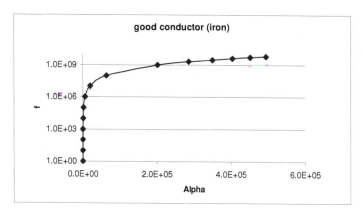

Fig. 3-10: Attenuation factor of iron for frequencies up to 6 GHz

Hui [22] investigated to see if a reflectivity minimisation of windowpanes is possible by changing the resistance of the layer. The result was, that this does not only affect the reflections at the coated surface of the glass, it also minimises the reflections from the uncoated side of the windowpane. Figure 3-11 shows the return losses dependent on the resistance of the coating.

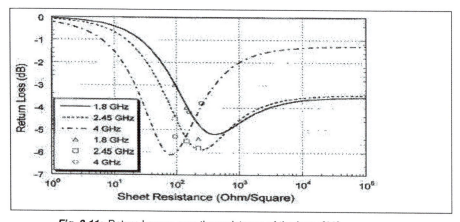

Fig. 3-11: Return Losses over the resistance of the layer [22]

Referring to Figure 3-11, it is apparent, that for a specific value of resistance, the reflectivity has its minimum. The value of the resistance for the lowest reflectivity decreases with increasing the frequency. When the resistance is very small, then the material may be presumed to be a perfect conductor, so the losses due to reflection are going to be maximum. Otherwise, if the resistance is very high, the return losses will almost be constant for any frequency.

Information about the frequency-dependence of the conductivity or of the resistance of materials is not usually readily available. Thus, the measurement results in this project will give a prediction about the changing of these material properties. This was commonly done by others already when taking measurements and the accuracy of the results was satisfying [23]. An approximation of the conductivity is possible with the equations in Table 3.1.

This special case of reflectivity is also not supported by the math model due to the complexity of the problem. But this effect, discovered by Hui, would deliver a good base of discussion, when the analysis of the real measurements differs from the simulation results.

3.5.1 Modified propagation constant

Consider two adjacent metallic layers with good conductor properties. From the mathematical point of view this case is more complex as following equations will illustrate. Not only the propagation factor is complex, also the phase constant becomes complex. Thus, the derivation of the angle of refraction is complex, which is presented now. The transmitted electric field of a horizontal polarised wave in a lossy medium is given by:

$$\mathrm{E}^{\mathrm{t}} = \hat{a}_y \cdot \mathrm{E}_0 \cdot \mathrm{T} \cdot e^{-(\alpha_2 + j\beta_2) \cdot (x \cdot \sin\theta_t + z \cdot \cos\theta_t)} \qquad [3.5.1]$$

The angle of the transmitted wave is given by Snell's law:

$$\gamma_1 \cdot \sin\theta_i = \gamma_2 \cdot \sin\theta_t \qquad [3.5.2]$$

It is apparent, that we will arrive at a complex angle. Thus, it is necessary to derive the transmission angle in another way. Referring to equation 3.3.25 we can write:

$$
\begin{aligned}
\cos\theta_t = \sqrt{1 - \sin^2\theta_t} &= \sqrt{1 - \left(\frac{j\beta_1}{\alpha_2 + j\beta_2}\right)^2 \cdot \sin^2\theta_i} \\
&= \sqrt{1 - \left(\frac{\beta_1\beta_2 + j\alpha_2\beta_1}{\alpha_2^2 + \beta_2^2}\right)^2 \cdot \sin^2\theta_i} \qquad [3.5.3] \\
&= \sqrt{1 - \frac{\beta_1^2(\beta_2^2 - \alpha_2^2)}{(\alpha_2^2 + \beta_2^2)^2} \cdot \sin^2\theta_i + j \cdot \frac{2 \cdot \alpha_2\beta_1^2\beta_2}{(\alpha_2^2 + \beta_2^2)^2} \cdot \sin^2\theta_i} = \sqrt{a + j \cdot b}
\end{aligned}
$$

The complex square root is abbreviated because of not losing track of things. The real part a, and the imaginary part b are given by:

$$a = 1 - \frac{\beta_1^2 (\beta_2^2 - \alpha_2^2)}{(\alpha_2^2 + \beta_2^2)^2} \cdot \sin^2 \theta_i$$

$$b = \frac{2 \cdot \alpha_2 \beta_1^2 \beta_2}{(\alpha_2^2 + \beta_2^2)^2} \cdot \sin^2 \theta_i$$

[3.5.4]

The next step is to extract the square root of this complex number. The result can be written as:

$$\sqrt[4]{a^2 + b^2} \cdot \left(\cos \left(\frac{\arctan\left(\frac{b}{a}\right) \cdot \frac{\pi}{180}}{2} \right) + j \cdot \sin \left(\frac{\arctan\left(\frac{b}{a}\right) \cdot \frac{\pi}{180}}{2} \right) \right) = \cos \theta_t$$

[3.5.5]

With the insertion of parameter s and ζ, equation 3.5.5 can be rewritten as:

$$\cos \theta_t = s(\cos \varsigma + j \cdot \sin \varsigma)$$

[3.5.6]

The parameter s refers to the root-expression. The other symbol, ζ, relates to the arguments of cosine and the sine – functions.

Equation 3.5.1 can now be rewritten as:

$$E' = E_0 \cdot T \cdot \exp\left\{ -(\alpha_2 + j \cdot \beta_2) \left[x \cdot \frac{j\beta_1}{\alpha_2 + j \cdot \beta_2} \cdot \sin \theta_i + z \cdot s (\cos \varsigma + j \cdot \sin \varsigma) \right] \right\}$$ [3.5.7]

Once again this equation will be decomposed in real and imaginary parts. Additionally, the following simplifications have been made:

$$p = s(\alpha_2 \cdot \cos \varsigma - \beta_2 \cdot \sin \varsigma) = \alpha_{2e}$$

$$q = s(\alpha_2 \cdot \sin \varsigma + \beta_2 \cdot \cos \varsigma)$$

[3.5.8]

The electric field is now given by:

$$E^t = E_0 \cdot T \cdot e^{-z \cdot p} \cdot \exp\{-j(\beta_1 \cdot x \cdot \sin\theta_i + z \cdot q)\} \qquad [3.5.9]$$

Assuming that E_0T is real, we can write for equation 3.5.9:

$$E^t = E_0 \cdot T \cdot e^{-z \cdot p} \cdot \cos(\omega t - (\beta_1 \cdot x \cdot \sin\theta_i + z \cdot q)) \qquad [3.5.10]$$

This is apparently not a uniform plane wave, whose electric and magnetic field have generally the form, as already presented with equations 2.2.24 and 2.2.25.

The angle of transmission, which can be described by the argument of the cosine function, is not the commonly utilised angle θ_t. This new angle is called ψ_2, which is not equal to θ_t. This new angle can be obtained with the following equation:

$$\sin\psi_2 = \frac{u}{\sqrt{u^2 + q^2}} \qquad [3.5.11]$$

where q is defined by equation 3.5.8 and u can be written as:

$$u = \beta_1 \cdot \sin\theta_i \qquad [3.5.12]$$

Finally the electric field is given by:

$$E^t = E_0 \cdot T \cdot e^{-z \cdot p} \operatorname{Re}(\exp[j \cdot (\omega t - \beta_{2e}(x \cdot \sin\psi_2 + z \cdot \cos\psi_2))] \qquad [3.5.13]$$

with: $\qquad\qquad \beta_{2e} = \sqrt{u^2 + q^2} \qquad\qquad\qquad [3.5.14]$

Figure 3-12 illustrates the utilised parameters. For a calculation of the phase velocity another derivation is also necessary [3].

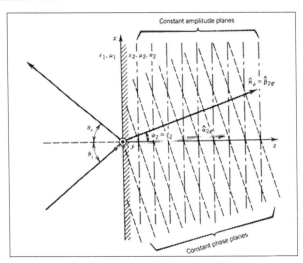

Fig. 3-12: Uniform plane wave impinging on a boundary of another medium [3]

Not only is the transmission angle changing, the propagation constant γ is also differing from the commonly used procedure. The magnitude of the difference between the intrinsic and the modified propagation constant depends on the ratio between the conductivities of both media. If the ratio reaches a certain limit, then it is absolutely necessary to utilise the modified propagation constant instead of the intrinsic one. The modified propagation constants are only used for non uniform plane waves. In the equations above, the expressions α_{2e} and β_{2e} are called the effective or modified attenuation and phase constant. Holmes and Balanis [24] derived a method to obtain a modified propagation constant and illustrated a graphical comparison between the modified and intrinsic propagation constants. Additionally they found out that when medium 2 is a perfect dielectric, the intrinsic and modified propagation constants show great differences for various incidence angles. Therefore, the modified parameters have to be utilised for the calculations. Feick [25] has used equations, derived from Balanis [24], for a software model, which simulates wave propagation through several construction materials. The modified propagation constants can be written as:

$$\alpha_{2e} = \sqrt{\frac{1}{2} \cdot \mathrm{Re}(\gamma^2) + \omega^2 \cdot \varepsilon_0 \mu_0 \cdot \sin^2 \theta_i + \left|\omega^2 \cdot \varepsilon_0 \mu_0 \cdot \sin^2 \theta_i - \gamma^2\right|}$$
[3.5.15]

$$\beta_{2e} = \sqrt{\frac{1}{2} \cdot (-\mathrm{Re}(\gamma^2)) + \omega^2 \cdot \varepsilon_0 \mu_0 \cdot \sin^2 \theta_i + \left|\omega^2 \cdot \varepsilon_0 \mu_0 \cdot \sin^2 \theta_i - \gamma^2\right|}$$
[3.5.16]

Figure 3-13 shows the difference between the intrinsic and the modified propagation constant. Both media are assumed to be conductive. For medium one were two different materials used with different conductivities, respectively. Their propagation constants are represented by the dashed lines. It is observable, that when the conductivities of both media are approximately equal it is sufficient to utilise the intrinsic parameters, because they are matching with the modified constants, which are denoted with the drawn through lines. Otherwise the discrepancies are too big, and the modified parameters have to be used.

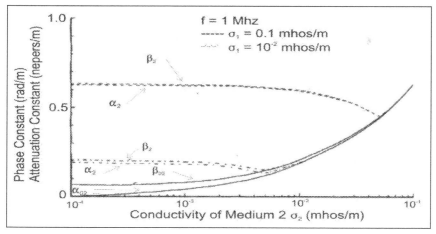

Fig. 3-13: Comparison between intrinsic and modified propagation constants [3]

These investigations could be very significant, when for instance two adjacent layers of a windowpane consist of conductive materials causing the EM waves to be affected in their propagation in another way as generally assumed.

This section showed how many considerations have to be made to develop an applicable model for window construction with conductive materials. It is apparent that the effort to realise a good approximation of EM wave propagation will increase when, for instance, a window with FSS [18, 19] will be investigated. Therefore, this preview is convenient for revealing the potential   complexity   of   this   work.

4. Experimental setup and procedure

Due to the radiation patterns of all the different utilised antenna types, their distance to the specimen and the propagation frequency dependence, especially in the VHF range, the measurement setup is constructed in the following way that it is applicable for the complete frequency range of interest. This frequency range consists of several radio services, which are interesting to investigate, because of their widespread use in our common life. Additionally, investigations in the X-Band will also be done. Military, such as aircraft radar, space research and broadcast services via satellite operate in the frequency range of 8 GHz – 12.5 GHz. In terms of this experiment the working frequencies of following services could be analysed:

service	Frequency
broadcasting services	87.5 MHz – 108 MHz
DVB-T (*Digital Video Broadcasting – Terrestrial*)	177.5 MHz – 803 MHz
ISM (*Industrial, Scientific, and Medical Band*)	2.4 GHz
Bluetooth	2.4 GHz
WLAN IEEE 802.11(b) standard (*Wireless Local Area Network*)	2.4 GHz
MobileWiMax (*Worldwide Interoperability Microwave Access*) (IEEE 802.16e-2005)	2.3 GHz – 2.7 GHz, 3.3 GHz – 3.4 GHz
X-Band services (*several military and satellite services*)	8 GHz – 12 GHz

Table 4.1: Overview of different services in the frequency range of interest

All measurements satisfy the defined minimum distance to achieve far-field characteristics for antenna measurements, so only a phase error of the EM waves is expected. Unfortunately, it is not possible to evaluate uniform plane waves, because in theory, the distance between transmitter and receiver has to be infinite. In the following sections, the setup is explained in more detail.

4.1. Measurement setup

The experiment takes place in a sufficiently long corridor in the cellar of the University of Applied Sciences and Arts in Hannover. The dimensions of the corridor are 2.1 metres in height and 3 metres in width. The ceiling, the ground and the side walls consist of thick blocks. This environment seems to be suitable despite the fact that the ceiling is covered with pipes and the width has not the desired size. Therefore, it is expected that measurements with transmitter-receiver distances of several meters, necessary in the VHF and UHF frequency ranges, will be disturbed by reflections from all the corridor bordered walls. On the other hand, this corridor is well shielded by thick walls, so that present broadcast services have no big influence on the experiment. The main task is to prevent the undesired reflections by the walls. Thus the area around the windowpane has to be completely shielded. Thus, a majority of EM waves is transmitted directly through the window and not by other propagation paths. A plot of the wall is illustrated in Figure 4-1:

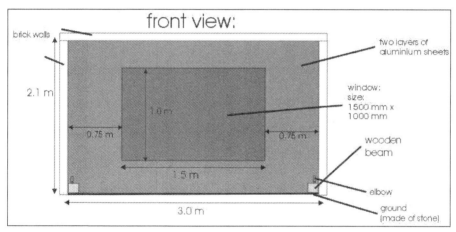

Fig. 4-1:　　　　　Front view of the specimen

The window-surrounding wall consists of plywood with a thickness of about 12 mm. Additionally, to prevent a transmission of EM waves through the wall, a polyurethane foam board, sandwiched by two sheets of aluminium with a thickness of about 1 mm each, provides good reflective characteristics. A test measurement for the investigation of the path losses through the plywood showed that the losses are approximately 20 dB higher in relation to the losses in free space. The side view in Figure 4-2 illustrates the setup of the wall. The wooden beams prevent the wall from falling down. The stillage is responsible for the robustness of the wall, and also for fixing the plywood with screws in it.

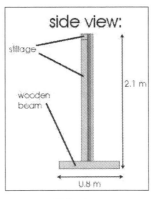

Fig. 4-2:   Side view

Two types of windows will be investigated with this setup. The dimensions are in both cases similar: Double-glazed windows, where both panes have a thickness of 4 mm with a gas-filled spacing of 12 mm, which is in total a complete window thickness of 2 cm. But the two windows' characteristics differ. The first window is a standard insulation float glass, and on the other hand the second one is a heat isolation float glass, which normally consists of metallic layers and inert gas-filled spacing.

The main problem of the wall is the weight of one of these modern windows, which is approximately around 50 kg. To change the window very easily and to support the weight, the construction in Figure 4-3 should be suitable.

The plywood ovorlaps tho pane at the top and the bottom by about 1 cm, so the window cannot fall forward. The wooden slats, which have the same thickness as the windowpane, sandwich the specimen. The strips of wood are fixed by screws with the slats, and also overlap the window, which prevents the window from falling backward. In Figure 4-4, showing the rear view, it is observable how the weight of the window could be supported.

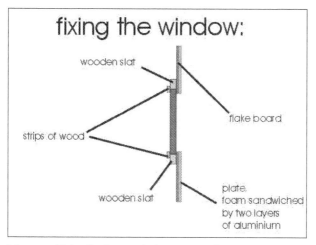

Fig. 4-3:   Fixing the heavy window construction

First of all a table holds the weight of the window. Directly on the tabletop lies the lower wooden beam, which is fixed by elbows on the left and right hand side of the stillage. Due to the table utilised, the height of the window is 86 cm over the ground.

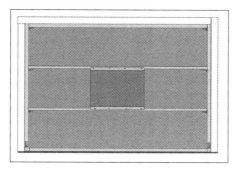

Fig. 4-4:   Rear view of the construction

The block diagram of the experimental setup for determining the path losses through the two different window types is presented in Figure 4-5, while the windowed wall is illustrated in Figure 4-6.

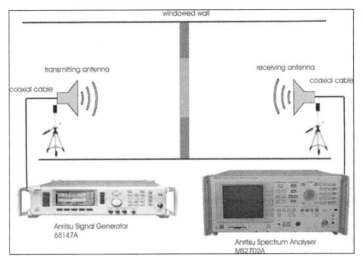

Fig. 4-5: Experimental setup for measuring the signal losses

Fig. 4-6: Windowed wall between transmitter and receiver

4.2. Utilised antennas

The wave propagation characteristics will be investigated in several frequency ranges where different types of antennas have to be utilised. For the VHF- frequencies (87.5 MHz – 108 MHz), a three-element Yagi antenna from Konni-Antennen and a biconical antenna (Rohde & Schwarz HK116 20 MHz – 300 MHz) will be applied.

Following pictures show these antennas:

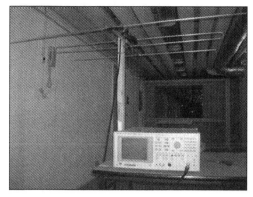

Fig. 4-7:   Three-element yagi antenna

The dimensions are as follows:

- length of folded dipole : 1.5 m
- length of reflector : 1.7 m
- length of director : 1.4 m
- length of boom : 0.85 m

Fig. 4-8: Biconical antenna HK116 (Rohde & Schwarz)

The measurements in the frequency range for DVB-T needs other additional antennas. A seven-element antenna from Konni-Antennen (470 MHz – 862 MHz) and a logarithmic periodical antenna (Rohde & Schwarz HL223 200 MHz – 1300 MHz) will be adopted. These antennas are illustrated in Figures 4-9 and 4-10:

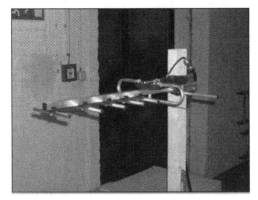

Fig. 4-9:   Seven - element - yagi antenna

The dimensions are as follows:

- length of folded dipole : 25 cm
- length of reflector : 30 cm
- length of the directors : 2 x 15.5 cm, 2 x 14.5 cm, 13.5 cm
- length of boom : 54 cm

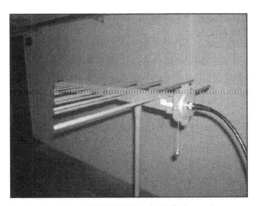

Fig. 4-10: Log. per. antenna HL223 (Rohde & Schwarz)

A convenient antenna for the frequencies from 1 GHz to 1.3 GHz is following illustrated log. per. antenna from Telefunken, which is utilised as transmitter and receiver antenna:

Fig. 4-11:   Log. per. antenna from Telefunken

Horn antennas are commonly utilised for frequencies above 1 GHz. For two possible measurement series, two identical horn antennas can be deployed, respectively. Two self constructed horn antennas from the Leibniz University of Hannover will be adopted for the frequency range from 2.2 GHz up to 3 GHz (Figure 4-12).

The signal losses in the X-Band (8 GHz – 12.5 GHz), are investigated with the horn antennas illustrated in Figure 4-13.

Fig. 4-12: Horn antenna for the frequency range from 2.2 GHz to 3 GHz

Aperture-dimensions:
- length : 12.8 cm
- heigth : 9.5 cm

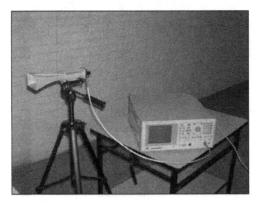

Fig. 4-13:   X-Band horn antenna

Aperture-dimensions:

- length : 7.7 cm
- height : 5.4 cm

4.3. Procedure

A lot of parameters are very significant to consider for detecting the correct signal losses through the window. First of all, the appropriate antennas in terms of the frequency have to be disposed. Furthermore, the alignment of the transmitting and receiving antenna has to be perfectly correct, which means that they must be fixed exactly facing each other.

To keep the conditions of the definition for far-field characteristics, the distance between the two antennas varies with type and frequency as well. The procedure in this investigation contains two measurement series. First of all, the losses of the transmission system, which means the wall in Figure 4-1 is included without an installed window, have to be measured in the whole frequency range. So, the characteristics of the transmission system are revealed, because the parameters of an antenna changes with the working frequency such as the gain, directivity, radiation characteristic and input impedance. But not only are the properties of the antennas registered, additionally, the attenuation of coaxial cables increases with frequency too, and this effect is considered as well.

With the second measurement series, the losses of the same unmodified experimental setup, but with an installed window are recorded. The differences of the detected signal strengths are obviously the losses caused by the window.

The recorded results in chapter 5 are based on this procedure. Unfortunately, it is not possible to adjust different incident angles of the signal on the specimen, because the desired accuracy of the traversable tripod is not precise enough. Therefore, an investigation in this direction cannot be done.

5. Measurement results

Several interesting frequency ranges have been taken into account in examining the signal losses when EM waves propagate through the two windowpanes briefly introduced in chapter 4. The results are illustrated in the following curves. In the first section of this chapter, the adaptability of the self-coded software will be demonstrated. Afterwards, the measured path losses, produced by passing the window construction will be shown. Tables of the results can be found in the appendix.

5.1. Software - test

A simple scenario was utilised to test the software model, and to see whether the algorithm works correctly or not. A couple of test series should prove the correct simulation of total transmission ($\Gamma = 0$), which is only possible for vertical polarisation. The Brewster – angle does not exist for horizontal polarised waves, which can also be shown. Rappaport [1] still presents the expected results and the program has to confirm it.

The following figures illustrate the absolute value of the reflection coefficient for both linear polarisations:

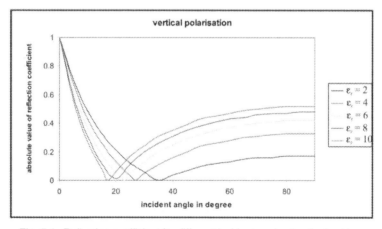

Fig. 5-1: Reflection coefficient for different incident angles (vertical pol.)

The *Brewster Angle* occurs when the reflection coefficient is zero, which is observable in Figure 5-1. The same simulation for horizontal polarisation provides the following results:

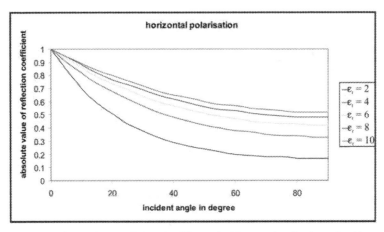

Fig. 5-2: Reflection coefficient for different incident angles (horizontal pol.)

As already mentioned, no *Brewster Angle* exists for horizontal polarisation and Figure 5-2 shows that the reflection coefficient never becomes zero.

The conclusion is that the software is able to calculate the transmission of EM waves at boundaries correctly, and therefore it is applicable to use it for the analysis of the real results, which are presented now.

5.2. 87.5 MHz - 108 MHz (radio broadcasting services)

Figure 5-3 presents the path losses for both windows in the frequency range of FM radio broadcasting services:

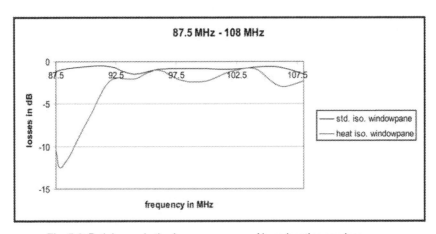

Fig. 5-3: Path losses in the frequency range of broadcasting services

5.3. 470 MHz - 860 MHz (DVB-T)

Unfortunately, the measurements of both window types were not possible to do under the same conditions. Despite this, the path losses for each windowpane are presented in one graph. However, no detailed comparison is meaningful.

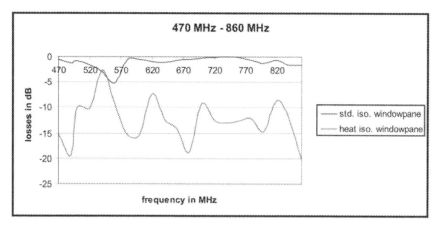

Fig. 5-4:   Path losses in the DVB-T frequency range

5.4. 1000 MHz - 1300 MHz

5.4.1 Horizontal polarisation

A pair of logarithmic periodical antennas, illustrated in Figure 4-11, were utilised to measure the signal losses through both window types in the frequency range of 1 GHz to 1.3 GHz. Figure 5-5 shows the results for horizontally polarised EM waves.

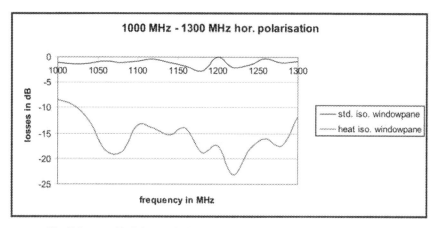

Fig. 5-5:   Path losses for horizontal polarisation

5.4.2 Vertical polarisation

Without changing the measurement conditions of the test series for horizontally polarised waves the same investigation were done for vertical polarisation. The results are plotted in Figure 5-6.

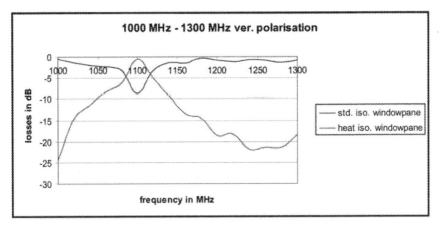

Fig. 5-6:   Path losses for vertical polarisation

5.5. 2.2 GHz - 3 GHz (WLAN, Bluetooth, WiMAX)

The measurements could not be performed under the same conditions, so an interpretation in detail between both windows is not expedient, but the trend of the curves is apparently observable.

These measurements were done with the in Figure 4-12 pictured horn antennas.

5.5.1 Horizontal polarisation

The results for horizontal polarisation are presented in Fig 5-7. The obtained results were measured with antennas which are not mounted on a tripod. Therefore inaccuracies are expected.

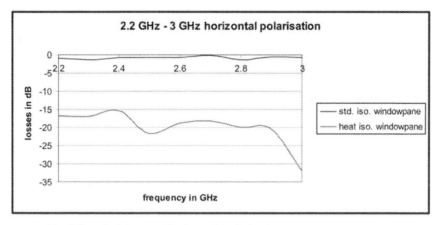

Fig. 5-7: Path losses of horizontally polarised waves

5.5.2 Vertical polarisation

The results for vertical polarisation are shown in Fig 5-8:

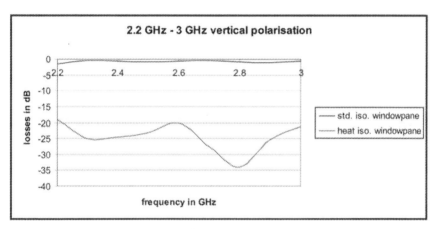

Fig. 5-8: Path losses of vertically polarised waves

5.6. 8 GHz - 12.5 GHz

The X-Band measurements are, in terms of new wireless communication systems very important. Two similar horn antennas, one of them is illustrated in Figure 4-13, were utilised to perform the investigation of signal losses in this frequency range. The measurement series include both linear polarisations.

5.6.1 Horizontal polarisation

The results for horizontal polarisation are illustrated in Figure 5-9:

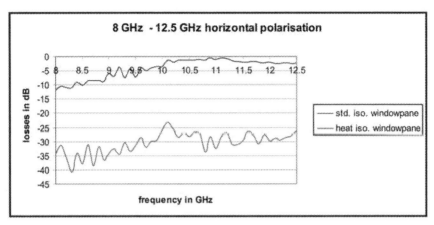

Fig. 5-9: Path losses of horizontally polarised waves

5.6.2 Vertical polarisation

Finally, the same measurements for vertically polarised EM waves are presented in Figure 5-10.

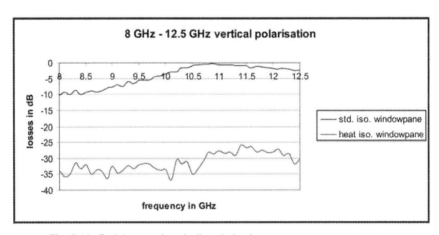

Fig. 5-10: Path losses of vertically polarised waves

6. Analysis

6.1 Overview over the complete frequency range

The results presented in chapter 5 show that the heat isolation windowpane has very strong signal losses in the whole upper frequency range. Only for the VHF – frequency range (Figure 5-3), are the losses comparatively in the same order of magnitude as the losses caused by the standard isolation window. Therefore, wireless services excepting radio broadcasting services, are highly disturbed by the heat isolation glass. Glass manufacturers include metallic coatings on the glass layers to achieve good heat isolation characteristics, but they did not consider the consequences for modern transmission systems. It is observable that the propagation losses increase with frequency. This is illustrated in Figure 6-1, where the trend of the signal losses over the whole frequency range investigated, is plotted for both window types. These results are measured relative to the losses incurred when no window was installed in the wall, pictured in Figure 4-6. Unfortunately, no convenient antennas for the frequency range between 3 GHz – 8.5 GHz were available. Therefore, this area in the plot is just an approximation, and will not be discussed because it could differ in reality.

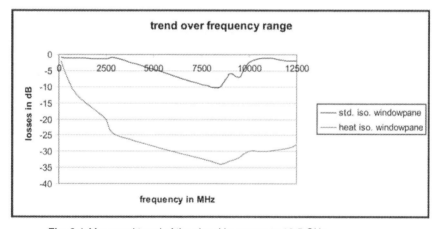

Fig. 6-1: Measured trend of the signal losses up to 12.5 GHz

The results from the standard isolation window can be confirmed by previous investigations such as the investigations of path losses through several building constructions, performed by the NIST (National Institute of Standards and Technology) Gaithersburg, Maryland [26]. They have measured the attenuation of single-glazed panes with different thicknesses for the frequencies from 0.5 GHz up to 2 GHz.

Their results are comparable to the in Figure 6-1 presented results:

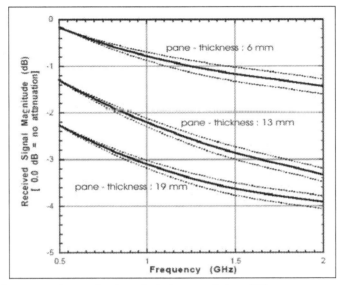

Fig. 6-2:   Attenuation through single pane windows [26]

The measured results of the attenuation through standard glass by the NIST match with the results of this work. Despite the fact that the investigations are done for different window constructions, a comparison is tolerable. If considering the sum of the thicknesses of both windowpanes, neglecting the spacing, the signal losses, which are between 1 – 2 dB, are in the same order as the signal attenuation measured by the NIST.

The standard isolation window has no extra components, thus no special propagation effects, such as bad transmission characteristics for a specific frequency range, should occur. But the losses increase quickly after about 8 GHz. Inaccurate measurements, unexpected or abnormal wave propagation could not be the reason for this result, because the horn antennas utilised were very close to the window pane (0.5 m). Therefore the transmission line between transmitter and receiver is very short and only obstructed by the window itself. Also, horn antennas have narrow lobes, resulting in a very good directivity.

The thickness of the window construction is 2 cm, and the wavelength of 8 GHz is 3.75 cm. Consider that the wavelength of an EM wave will get longer when propagating into a medium with lower wave impedance. It could be possible that the losses in this range, related to the remaining frequency range, are quite high due to superposition effects inside of the window, because the wavelength of the signal is about two times larger than the thickness of the window construction.

The incident angle of the EM waves is normal, so reflected waves will propagate directly back to the signal source and influence new radiated waves coming from the transmitting antenna. Note that multiple reflections occur, because the complete window has four boundaries and reflected waves will be generated on each of them. Destructive superposition of this cluster of EM waves would effect the overall transmission of energy through the window. Therefore, the transmitted energy in this frequency range may be very small.

The reason for the high losses of heat isolation windows, represented by the red line, is the special window setup. To achieve an energy-efficient effect, layers and inert gas-filled spacing are implemented in a window construction. The thickness of the layers varies from a very tiny order of about 10 nm up to 0.4 µm [27]. The utilised materials are silver and tin oxide, both of which have good conductive properties. The spacing is gas-filled with air (but very seldom), argon, krypton, xenon, CO_2 and sulphur hexafluoride.

Consequently, the effect around 8 GHz, which was observable for the case of the standard isolation window, should also occur for the heat isolation window due to the same dimensions, and the trend is quite similar. Apparently, the added materials are only responsible for the higher losses, because conductive materials produce high reflectivity on EM waves, which results in poor transmission characteristics. However, they do not affect the EM waves at specific frequencies. The reason is that the metallic coatings are regular and do not have a special design as the examples already introduced in chapter 2.

Unfortunately, it is only speculation, why the wave propagation in heat isolation windows is very bad, because the exact setup of this window is unknown. Therefore, the basis of the assumptions made relating to the window designs, is information published by famous glass manufacturers [28].

The wireless systems of the future, obviously working at higher frequencies, will have enormous problems to provide indoor receivers with sufficient field strength. As mentioned briefly in chapter 1, the outdoor-to-indoor wave propagation works well through windows, while for instance concrete, or masonry block walls strongly attenuate the signals propagating through them [26]. Thus, these investigations are interesting in terms of predicting wave transmission through windows.

6.2 Discussions about the results in detail

6.2.1 VHF - frequencies

The results in the lowest investigated frequency–range show constant losses around 1 dB in the standard isolation window and an average loss of about 2 dB caused by the heat isolation window. In the latter case, very high losses were measured for the lowest registered frequencies. It is improbable that a certain effect could occur, which makes it impossible for EM waves to penetrate the window. Therefore, these must be inaccurate measurements. Perhaps, the windowed wall was unintentionally modified by the small lockable passage at the exterior of the wall and EM waves, reflected at the side walls, could pass through a small gap and destructively superimpose with waves transmitted through the window. There are several interpretations possible. The measurement setup is similar to the previous test run with the isolation windowpane, because the antenna positions and distances were fixed. It is important to note, that especially for the frequencies, where the distance between transmitter and receiver are quite large, small alterations of the wall could influence the accuracy of the measurements. The hallway – environment, which was good, but not perfect for the investigations, forces one to conduct the experiment in that way. Also a small gap at the ceiling was permanently present. These considerations have to be taken into account for all measurements up to 2 GHz. Altogether, the tendency is clearly observable and the conclusion is that the losses are very tiny and no distinction between both window types is necessary.

6.2.2 UHF - frequencies

Figure 5-4 illustrates two curves, which are not recorded under the same conditions. To discuss the results, it is convenient to go in detail with separated examinations of each curve. Figures 6-3 and 6-4 show all recorded measurement series to obtain the final results in Figure 5-4:

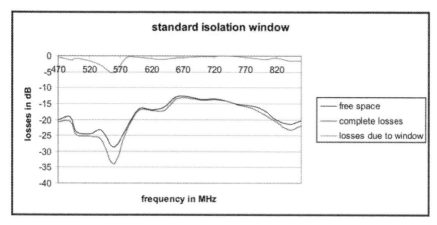

Fig. 6-3:   Standard isolation window

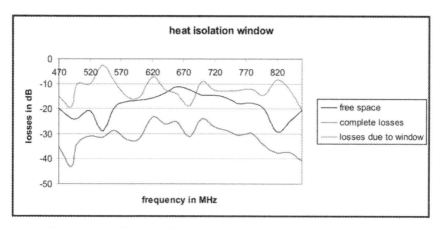

Fig. 6-4:   Heat isolation window

In both pictures, the blue line denotes the free space losses, while the red line stands for the losses with a window. The green line is the difference between the other two, and therefore, the losses produced by the window. The analogy of both series is observable when comparing the free space losses of the transmission system. The trend of the blue line matches in both figures, because at approximately 550 MHz and 820 MHz, maximum losses occur. The transmission through the standard isolation window is as good as in 6.2.1. On the other hand, Figure 6-4 illustrates that not only the losses are higher; also considerable fluctuations have been recorded.

6.2.3 1 GHz - 1.3 GHz

The measurement with the antennas utilised was a challenge, because it was very difficult to fasten them on a mast and align them perfectly too. But an interesting effect was observable.

The test run with horizontal polarised EM waves was not so spectacular, despite that the losses of the standard isolation window were very close to 0 dB. This is impossible, because no total transmission of horizontal polarised waves is possible. Therefore, this must be inaccurate. However, the standard isolation glass shows good transmission characteristics, while the same fluctuations as already described in 6.2.2 are illustrated for the heat isolation window. But the result for vertical polarised EM waves gives a good basis for discussion. Unusual, when comparing with the results already presented is that the heat isolation window has a better transmission characteristic than the standard isolation window at the frequency of 1.1 GHz (Figure 5-6). Additionally, the transmission of EM waves at the same frequency is close to perfect. Furthermore, the standard windowpane, which usually has very small losses, suddenly shows the worst detected transmission characteristic, at exactly 1.1 GHz. The conditions were the same for both series, so the assumption that this could be a total transmission of EM waves, is legitimate. Considering the high reflective metallic coatings of the window, it is unlikely one may presume this could be total transmission. Nevertheless, this has to be checked by additional measurements, which were in terms of this investigation hard to realise, because modifying the setup, in such a way that the conditions do not change, is impossible. As already mentioned, the losses of the standard isolation window for this frequency are quite high. This is as abnormal as the total transmission of the heat isolation window, and has to be checked as well.

6.2.4 2.2 GHz - 3GHz

Figures 5-7 and 5-8 show the signal losses for both linear polarisations. The transmission characteristics of the standard isolation window are unchanged, while the other window does not provide good transmission properties. The losses vary from 15 dB up to 35 dB, which denotes that the maximum transmitted power is only 3 % of the origin signal strength. That value is very bad, because a lot of important broadcasting services are working in this frequency range. Therefore, it is not a good advice to have heat isolating windows, and similarly, wireless computer networks such as WLAN or Bluetooth.

6.2.5 8 GHz - 12.5 GHz

The unexpectedly high losses of standard isolation windowpanes have already been discussed in section 6.1. Nevertheless, the transmission characteristic is as good as in the other investigated frequency ranges.

Additionally, the measured results for the X-Band are shown in Figures 5-9 and 5-10. The coated window only allows less than 0.1 % of energy to be transmitted through its construction.

The conclusion is quite simple: Wireless systems of the next generation will be strongly affected by objects, when propagating through them. Therefore, in our case, windows need to have an alternative setup, instead of the current one to provide a smooth signal transmission. The only possibility is to find other options to achieve the same characteristics that heat isolation windows have. The utilisation of metallic coatings will produce too big problems for the outdoor to indoor wave propagation, and it is definitely a problem in the next few years.

6.3. Software model

Chapter 5 presented in terms of total transmission the correct functionality of the software model. Because no conductive materials are supported, it is only appropriate to compare the measured results of standard isolation windows with the results of the model. The algorithm is briefly described in Chapter 3.4, and the source code is attached in the appendix. The measurements, described in the previous sections, are only made for normally incident EM waves.

In this simulation, several parameters were varied to obtain a meaningful result. Every line in Figure 6-5 represents a separate angle of incidence. The value of the permittivity, allocated to the glass layers, is the modifiable parameter in this simulation. Assisted by the equations for the *transmission* and *reflection coefficient* (equations 3.3.26, 3.3.27, 3.3.45, 3.3.46), and the simulation results presented in section 5.1, it is expected that the losses will increase with higher permittivity and with increasing angle of incidence.

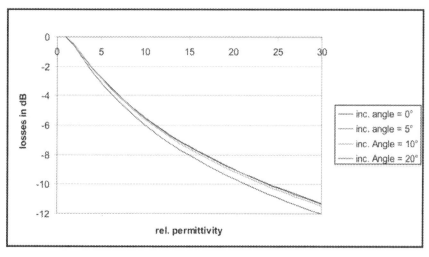

Fig. 6-5:   Losses in relation to the permittivity

The increase of losses by increasing the permittivity can be illustrated by the following picture:

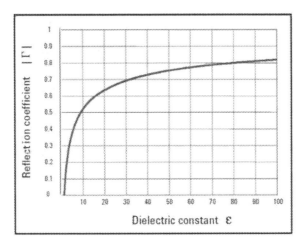

Fig. 6-6: Reflection coefficient versus dielectric constant [29]

The wave propagation through windows is essentially determined by reflection effects. Figure 6-6 illustrates the absolute magnitude of the reflection coefficient in relation to the permittivity. It is shown that the amount of reflected energy increases with permittivity. Therefore, the simulation results confirm this behaviour. Furthermore, higher incident angles increase the losses as well, which is apparently not the most important parameter for the losses, because the difference between the curves is very small for small incident angles. This is observable by analysing equation 3.3.28 or 3.3.47. The equation for the transmitted power density is determined by trigonometric functions. For small incident angles, the sine function has a bigger gradient than the cosine, and therefore, the sine argument determines the changing path losses. The difference between the curves is getting higher when the incident angle becomes larger, but this trend changes when the decrease of the cosine function overbalances the increase of the sine.

Finally, a comparison with the results of the standard isolation window is made. It was not possible to align the antennas perfectly. Thus, it is not sure that the EM waves impinge normally on the window. But the incident angle is definitely only in the order of a few degrees, so the red line in Figure 6-5 should be the worst case for the analysis.

In literature glass has a permittivity around four, a negligible loss factor and very low conductivity. Figure 6-7 shows several examples of different glass types.

Material	Parameter	Parallel Plate	Split Cavity	60 mm TE$_{01}$	Fabry Perot	Reentrant Cavity
Corning 7980 (batch 34604)	f	5.07	8.86	9.64	60	
	ε'	3.848	3.792	3.826	3.858	
	$\tan\delta$	6.85×10^{-5}	4.91×10^{-4}	1.4×10^{-4}	1.70×10^{-4}	
Corning 7980 (batch 34605)	f	5.07	8.86	9.63	58.7	
	ε'	3.844	3.796	3.826	3.843	
	$\tan\delta$	6.8×10^{-5}	1.48×10^{-4}	1.4×10^{-4}	5.45×10^{-4}	
Corning 7980 (batch 34606)	f	5.26	8.86	9.64	58.7	
	ε'	3.844	3.799	3.826	3.832	
	$\tan\delta$	6.90×10^{-5}	1.80×10^{-4}	1.3×10^{-4}	4.07×10^{-4}	
Corning 7940 (batch 44608)	f	4.89	8.86	9.64	59.9	
	ε'	3.847	3.808	3.826	3.781	
	$\tan\delta$	6.83×10^{-5}	1.77×10^{-4}	1.4×10^{-4}	5.09×10^{-4}	
Corning 1723 H. Bussey (1959)	f	9				1
	ε'	6.20				6.21
	$\tan\delta$	5.36×10^{-3}				3.2×10^{-3}

Fig. 6-7: Characterisation of different glass types [21]

Consider that the parameters are obtained by different measurement procedures. Despite that, the results do not differ in its value very much, even for different frequencies, so it is a basis for further discussions.

The permittivity, responsible for the reflectivity of glass (Figure 6-6), changes in a negligible order, thus the constant signal losses over the whole frequency range can be confirmed by Figure 6-7. Additionally, the loss factor is very small over the frequency range of interest. Therefore, as per definition, glass can be understood to not be a lossy material [21].

When comparing the simulation results with the measured values, it is noticeable that no substantial discrepancies are recognisable. Therefore, this comparison shows that the measured results seem to be valid. And, in addition, it also underlines the merits of the developed theoretical software model. Furthermore, the recorded losses of the heat isolation window are representative too, due to the measurement conditions not being modified in most cases.

7. Conclusions

This work presented the diversity of possibilities in the design of windows, and it also demonstrated on the basis of a test series the signal losses incurred in two, for construction purposes, commonly utilised window types. To confirm the obtained results, a software model was developed. In view of this model, the principles of EM wave propagation are introduced, which are subjected to the physics of optics. Based on the approved electromagnetic theory of Maxwell, a derivation of the mathematical definition of the simplest form of EM waves, the TEM wave, was presented. Considering the setup of modern window constructions, the mathematical description of EM wave propagation through different media is explained in detail. Reflection of propagating waves at windows, was also elaborated. Metallic coatings in windows produce more complex considerations in terms of wave propagation, and were briefly introduced, but not considered in the software model.

The practical part of this work shows the constructed measurement setup, the equipment utilised, and the measurements accomplished with their results. These results were analysed, discussed and compared with the theoretical software model.

From the electrical and practical point of view, glass facades, consisting of glass with no extra additions show very good transmission characteristics for the whole investigated frequency range. Referring to the available data of the material properties of glass, it is assumable that the amount of energy, which will be reflected by windows, should also be small for frequencies, higher than the frequency range investigated in terms of this thesis. Therefore, signals of wireless broadcasting services of the future, working in a frequency range where the EM waves will be very sensitive, should not have problems to propagate through a standard window.

On the other hand, the utilised metallic coatings influence EM waves enormously. The investigated heat isolation window is not applicable for already existing broadcasting services, and of course not applicable for services of the next generation. The amount of energy, which will be reflected, is quite high. Furthermore, the penetrated energy will be strongly attenuated by the conductive material.

Unfortunately, the glass manufacturers do not consider that their common window setup for providing certain special effects in terms of isolation and protection, affects electromagnetic negatively. A recommendation to the manufacturers is quite simple. They have to find alternative methods to achieve similar features without metallic coatings. But this may not happen. A compromise could be the FSS window system, introduced in Chapter 3 [18]. It should be possible to produce regular meshed layers, which let selective frequencies pass through it, and also providing the same comparable isolation or protection characteristics as the current window constructions. The big disadvantage is that not all broadcasting services will be supported by a single FSS. Widenberg described, that it is possible to design the FSS in that way, that it would let EM waves with several frequencies pass through it. But how many wireless systems should be considered from just one FSS? Consequently, the complexity and effort of the FSS design will increase, until reaching the physical limits. Therefore, supporting all broadcasting services is not realisable.

Sometimes it is desired that EM waves could propagate through windows, otherwise they need to be blocked. A window, absorbing or reflecting EM waves is also presented in chapter 3.

To do a safe job, the classical model of outdoor-to-indoor propagation should be utilised. Everybody wants to have an energy-saving house, thus these windows are commonly desired. Furthermore, wireless systems become important in everyone's life, so nobody would accept an indoor environment without broadcasting services. The conclusion is, installing a receiver at a convenient position outside of the building, amplifying the signal, and sending it to suitably placed indoor antennas, which provides sufficient field strength in the entire building.

Appendix:

A1: Symbols

VHF	Very – High – Frequency
UHF	Ultra – High – Frequency
L – Band	0.39 GHz – 1.55 GHz
S – Band	2 GHz – 4 GHz
C Band	4 GHz 8 GHz
X – Band	8 GHz – 12.5 GHz
K_a – Band	26.5 GHz – 40 GHz
UV	Ultraviolet
FM	Frequency modulation
GSM	Global System for Mobile Communications
UMTS	Universal Mobile Telecommunications System
DVB-T	Digital Video Broadcasting – Terrestrial
ISM	Industrial, Scientific, and Medical Band
WLAN	Wireless Local Area Network
WiMAX	Worldwide Interoperability Microwave
EM	electromagnetic
TEM	transverse electromagnetic mode
LOS	line – of – sight
FSS	Frequency Selective Surfaces
NIST	National Institute of Standards and Technology

$\hat{a}_i$	vector with unity length pointing in i-direction
n	refraction index
λ	wavelength
ω	angular frequency
f	frequency
v_P	phase velocity
c_o	speed of light (3×10^8 m/s)
ε	permittivity
ε_0	permittivity of vacuum (8.854×10^{-12} As/Vm)
ε_r	relative permittivity
ε'	real part of complex permittivity

ε''	imaginary part of permittivity
$\tan \delta$	loss tangent
μ	permeability
μ_0	permeability of vacuum ($4\pi \times 10^{-7}$ Wb/Am)
μ_r	relative permeability
σ	conductivity
θ_i	angle of incidence
θ_r	angle of reflection
θ_t	angle of transmission
θ_c	critical angle of total internal reflection
θ_b	Brewster angle (angle of total transmission)
θ_n	angle in polar coordinates of the complex intrinsic wave impedance
θ^+	positive phase difference from zero phase
θ^-	negative phase difference from zero phase
α	attenuation factor
α_{2e}	modified attenuation factor
β	phase constant
β_{2e}	modified phase constant
γ	propagation constant
γ^+	propagation constant in positive direction
γ^-	propagation constant in negative direction
E	electric field intensity
$\hat{E}_x$	total electric field in x direction
$\hat{E}^+$	electric field which describes a travelling wave in positive direction
$\hat{E}^-$	electric field which describes a travelling wave in negative direction
E_i	electric field of an incident EM wave
E_r	electric field of a reflected EM wave
E_t	electric field of a transmitted EM wave
D	electric flux density
H	magnetic field intensity
$\hat{H}^+$	magnetic field which describes a travelling wave in positive direction
$\hat{H}^-$	magnetic field which describes a travelling wave in negative direction
$\hat{H}_y$	total electric field in y direction
H_i	magnetic field of an incident EM wave
H_r	magnetic field of a reflected EM wave
H_t	magnetic field of a transmitted EM wave
B	magnetic flux density

J	current density
ρ	charge density
S	Poynting vector
S_{av}	average power density
$\Gamma_\perp$	reflection coefficient of a horizontal polarised EM wave
$T_\perp$	transmission coefficient of a horizontal polarised EM wave
$\Gamma_\parallel$	reflection coefficient of a vertical polarised EM wave
$T_\parallel$	transmission coefficient of a vertical polarised EM wave
δ	skin depth
j	imaginary part of a complex number
e	e function
Z_0	wave impedance
Z_{FS}	wave impedance of vacuum
η	characteristic wave impedance
S_R	power density at the receiving antenna
P_T	radiated energy by the transmitting antenna
G_T	gain of the transmitting antenna
R	distance transmitter – receiver

A2: Simulation source code

```
/************************************************************/
/*  WAVE PROPAGATION THROUGH WINDOW CONSTRUCTIONS V 1.3  */
/*                                                          */
/* Date: 21.06.06                                          */
/* Author : Nils Knauer                                    */
/*                                                          */
/* University of Applied Sciences and Arts Hannover (FH Hannover) / Germany  */
/* Waterford Institute of Technology (WIT) / Ireland      */
/*                                                          */
/************************************************************/

#include <iostream.h>
#include <stdio.h>
#include <math.h>
#include <stdlib.h>

// Definition of constant parameters

#define PI 3.141593
#define EPSILON_0 8.854e-12     // permittivity of vacuum
#define MU_0 1.257e-6           // permeability of vacuum
#define L_SPEED 3e+008          // speed of light in vaccuum
#define CONDUCTIVITY 1e-12   // conductivity
//----------------------------------------------------------------------------
// functions

        int input();
        int constants_calc();
        int refract();
        int result();
//----------------------------------------------------------------------------
// global parameters

        bool note = false;

        char polarisation;

        float layer_param[3][4];
        float incident_angle;
        float frequency;

        double wave_pow = 1;
        double layer_value[3][3];
        double omega;
        double init_pow_density;
        double pow_dens;
```

```
//-------------------------------------------------------------------------

int main()
{
        bool key;

        printf("WavePropagation V1.3 by Nils Knauer\n");

        input();
        constants_calc();
        refract();
        result();

        if (note == false)
        {
printf("The losses are so high that no energy is transmitted through  the window\n");
        }
                printf("Press any key!\n");
                scanf("%s",&key);

        return 0;
}
//************************************************************
int input()
{
        bool check = false;
        char data;

//        3 Layers : Layer1 : thickness : 4 mm, rel permeability : 1
//        conductivity : 1e-12
//        Layer2 : thickness : 12mm, rel permeability : 1
//                conductivity : 1e-12
//        Layer3 : the same as Layer 1

        layer_param[0][0] = 4;
        layer_param[0][2] = 1;
        layer_param[0][3] = (float)CONDUCTIVITY;
        layer_param[1][0] = 12;
        layer_param[1][2] = 1;
        layer_param[1][3] = (float)CONDUCTIVITY;
        layer_param[2][0] = 4;
        layer_param[2][2] = 1;
        layer_param[2][3] = (float)CONDUCTIVITY;

        do
        {
                frequency = 1000;
                printf("\nPlease type in the angle of incidence: ");
                scanf("%f",&incident_angle);
                printf("\nWhich polarisation ? ('h' or 'v'): ");
```

```c
            scanf("%s",&polarisation);
            printf("\nPlease type in the relative permittivity of layer 1: ");
            scanf("%f",&layer_param[0][1]);
            printf("\nPlease type in the relative permittivity of layer 2: ");
            scanf("%f",&layer_param[1][1]);
            printf("\nPlease type in the relative permittivity of layer 3: ");
            scanf("%f",&layer_param[2][1]);
            printf("\nInput data is correct? ('y' for yes, 'n' for no): ");
            scanf("%s",&data);

            if (data == 'y')
            {
                    check = true;
            }
            else
            {
                    check = false;
            }

    }while(check == false);

    return 0;

}
//*************************************************************/
int constants_calc()
{

// Calculation of all required parameters for each layer. It is assumed, that each
// layer is a good dielectric

    int i;

    omega = 2 * PI * frequency * 1e+06;

    for (i=0; i <= 2; i++)
    {
            // phase constant
            layer_value[i][0] = omega * sqrt(MU_0 * layer_param[i][2]
    * EPSILON_0 * layer_param[i][1]);
            // phase velocity
            layer_value[i][1] = 1 / (sqrt(MU_0 * layer_param[i][2]
    * EPSILON_0 * layer_param[i][1]));
            // intrinsic impedance
            layer_value[i][2] = sqrt((MU_0 * layer_param[i][2])
    / (EPSILON_0 * layer_param[i][1]));
    }


    if (polarisation == 'h')
    {
```

```c
            init_pow_density = (sin((PI/180) * incident_angle)
+ cos((PI/180) * incident_angle))
                                    / (2 * sqrt(MU_0 / EPSILON_0));
        }

        if (polarisation == 'v')
        {
            init_pow_density = (cos((PI/180) * incident_angle) - sin((PI/180)
* incident_angle))/ (2 * sqrt(MU_0
/ EPSILON_0));
        }

//      printf("init_pow_density:%f\n",init_pow_density);

        return 0;

}
/****************************************************************/
int refract()
{
        int i;

        double r_coeff, t_coeff, help;

// horizontal polarisation

        if (polarisation == 'h')
        {
                for (i=0; i<=2; i++)
                {
                        if (i == 0)
                        {
                                t_coeff = (2 * cos((PI/180) * incident_angle))
                                                / (cos((PI/180) * incident_angle)
                                                + sqrt(layer_param[i][1])
                                                * sqrt(1 - (1 / layer_param[i][1])
                                                * sin((PI/180) * incident_angle)
* sin((PI/180)
* incident_angle)));

                                incident_angle = (float)(180/PI
* asin((omega * sqrt(MU_0
* EPSILON_0) / layer_value[i][0])
                                                * sin((PI/180) * incident_angle)));

                                r_coeff = t_coeff - 1;

                        }

                        else
                        {
```

```
                        t_coeff = (2 * cos((PI/180) * incident_angle))
        / (cos((PI/180) * incident_angle)
        + sqrt(layer_param[i][1] / layer_param[i-1][1]))
* sqrt(1 - (layer_param[i-1][1]
        / layer_param[i][1]) * sin((PI/180)
        * incident_angle) * sin((PI/180)
        * incident_angle)));

                        incident_angle = (float)(180/PI
        * asin((layer_value[i-1][0]
                                / layer_value[i][0]) * sin((PI/180)
        * incident_angle)));

                        r_coeff = t_coeff - 1;


                }

                pow_dens = (wave_pow * wave_pow * t_coeff * t_coeff)
        / (2 * layer_value[i][2]) * (sin((PI/180)
        * incident_angle) + cos((PI/180)
        * incident_angle));

                wave_pow = wave_pow * t_coeff;
            }

        t_coeff = (2 * cos((PI/180) * incident_angle))
                                / (cos((PI/180) * incident_angle)
                                + sqrt(1 / layer_param[2][1])
                                * sqrt(1 - (layer_param[2][1])
                                * sin((PI/180) * incident_angle)
                                * sin((PI/180) * incident_angle)));

        incident_angle = (float)(180/PI
        * asin((layer_value[2][0] / (omega * sqrt(MU_0
        * EPSILON_0))) * sin((PI/180) * incident_angle)));

                r_coeff = t_coeff - 1;

                pow_dens = (wave_pow * wave_pow * t_coeff * t_coeff)
        / (2 * (120*PI)) * (sin((PI/180) * incident_angle)
        + cos((PI/180) * incident_angle));

                wave_pow = wave_pow * t_coeff;
        }

// vertical polarisation

        if (polarisation == 'v')
        {
```

```c
for (i=0; i<=2; i++)
{
        if (i == 0)
        {
                help = 1 - (1 / layer_param[i][1])
                                * sin((PI/180) * incident_angle)
                                * sin((PI/180) * incident_angle);

                t_coeff = (2 * sqrt(1 / layer_param[i][1])
                                * cos((PI/180) * incident_angle))
                                / (cos((PI/180) * incident_angle)
                                + sqrt(1 / layer_param[i][1])
                                * sqrt(help));

                incident_angle = (float)(180/PI
* asin((omega * sqrt(MU_0
* EPSILON_0) / layer_value[i][0])
* sin((PI/180) * incident_angle)));

                r_coeff = 1 - ( (120*PI) / layer_value[i][2] ) * t_coeff;


        }
        else
        {
                help = 1 - ((layer_param[i-1][1] / layer_param[i][1])
                                * sin((PI/180) * incident_angle)
                                * sin((PI/180) * incident_angle));

                t_coeff = (2 * sqrt(layer_param[i-1][1]
                                / layer_param[i][1])* cos((PI/180)
* incident_angle)) / (cos((PI/180)
* incident_angle) + sqrt(layer_param[i-1][1]
/ layer_param[i][1]) * sqrt(help));

                incident_angle = (float)(180/PI
* asin((layer_value[i-1][0]
                                / layer_value[i][0]) * sin((PI/180)
* incident_angle)));

                r_coeff = 1 - (layer_value[i-1][2]
                / layer_value[i][2]) * t_coeff;

        }

        pow_dens = (wave_pow * wave_pow * t_coeff * t_coeff)
                        / (2 * layer_value[i][2]) * (sin((PI/180)
* incident_angle)
                        + cos((PI/180) * incident_angle));
```

```c
                              wave_pow = wave_pow * t_coeff;
                       }

              help = 1 - ((layer_param[2][1])
                              * sin((PI/180) * incident_angle)
                              * sin((PI/180) * incident_angle));

              t_coeff = (2 * cos((PI/180) * incident_angle)
                      * sqrt(layer_param[2][1]))/ (cos((PI/180) * incident_angle)
                      + sqrt(layer_param[2][1]) * sqrt(help));

              incident_angle = (float)(180/PI * asin((layer_value[2][0]
                              / (omega * sqrt(MU_0 * EPSILON_0)))
                              * sin((PI/180) * incident_angle)));

              r_coeff = 1 - (layer_value[2][2] / (120*PI)) * t_coeff;

              pow_dens = (wave_pow * wave_pow * t_coeff * t_coeff)
       / (2 * (120*PI) * (sin((PI/180) * incident_angle)
       +  cos((PI/180) * incident_angle));

              wave_pow = wave_pow * t_coeff;
       }

       return 0;
}
//***********************************************************/
int result()
{

       float power_density;

       power_density = 10 * (float)log10(pow_dens / init_pow_density);

       if (pow_dens != 0)
       {
printf("angle: %.2f Path losses:%.2f dB;\n",incident_angle,  power_density);
              note = true;
       }

       return 0;
```

A3: Measurement data

1. 87.5 MHz – 108 MHz:

frequency in MHz	free-space loss in dB	standard iso in dB	Δ in dB	heat – iso in dB	Δ in dB
87.5	-16.9	-18.1	-1.2	-27.4	-10.5
88.0	-15.2	-16.1	-0.9	-27.6	-12.4
90	-15.4	-16	-0.6	-22.6	-7.2
92	-13.2	-13.8	-0.6	-15.5	-2.3
94	-13.3	-14.8	-1.5	-15.4	-2.1
96	-15	-15.9	-0.9	-16	-1
98	-14.3	-15.1	-0.8	-16.5	-2.2
100	-18.8	-19.6	-0.8	-21.1	-2.3
102	-25.9	-26.8	-0.9	-27.1	-1.2
104	-27.2	-27.9	-0.7	-28	-0.8
106	-20.5	-21.2	-0.7	-23.4	-2.9
108	-25.3	-26.8	-1.5	-27.6	-2.3

2. 470 MHz – 862 MHz (standard isolation window):

frequency in MHz	free-space loss in dB	standard iso in dB	Δ in dB
470	-20.18	-20.68	-0.5
490	-19.21	-20.45	-1.24
500	-23.89	-24.7	-0.81
520	-24.44	-25.23	1.6
540	-23.52	-26.59	-3.07
560	-28.72	-33.94	-5.22
580	-23.03	-23.76	-0.73
600	-16.81	-17.34	-0.53
620	-16.91	-17.18	-0.99
640	-16.24	-17.33	-1.09
660	-12.87	-13.53	-0.66
680	-12.72	-13.24	-0.52
700	-13.67	-13.94	-0.27
720	-13.49	-13.67	-0.18
740	-14.11	-14.11	0
760	-15.28	-15.4	-0.12
780	-15.62	-16.26	-0.64
800	-17.12	-18.33	-1.21
820	-20.08	-20.72	-0.64
840	-21.51	-23.15	-1.64
860	-20.38	-21.98	-1.6

3. 470 MHz – 862 MHz (heat isolation window):

frequency in MHz	free-space loss in dB	heat – iso in dB	Δ in dB
470	-19.9	-34.98	-15.08
490	-23.8	-43.27	-19.47
500	-23.8	-33.87	-10.07
520	-20.7	-30.85	-10.15
540	-28.9	-31.53	-2.63
560	-19.5	-28.71	-9.21
580	-17.22	-32.39	-15.17
600	-16.65	-32.12	-15.47
620	-15.87	-23.18	-7.31
640	-13.81	-26.05	-12.24
660	-11.22	-25.34	-14.12
680	-12.29	-31.08	-18.79
700	-14.54	-23.8	-9.26
720	-14.63	-27.21	-12.58
740	-15.78	-28.73	-12.95
760	-18.02	-30.52	-12.5
780	-17.73	-29.89	-12.16
800	-19.96	-34.61	-14.65
820	-29.15	-37.65	-8.5
840	-25.06	-37.34	-12.28
860	-20.53	-40.75	-20.22

4. 1 GHz – 1.3 GHz horizontal polarisation:

frequency in MHz	free-space loss in dB	standard iso in dB	Δ in dB	heat – iso in dB	Δ in dB
1000	-19.56	-20.72	-1.16	-27.99	-8.43
1020	-21.8	-23.23	-1.43	-31.33	-9.53
1040	-16.29	-17.56	-1.27	-29.04	-12.75
1060	-12.32	-13.12	-0.8	-30.7	-18.38
1080	-11.03	-12.12	-1.09	-29.73	-18.7
1100	-11.2	-12.02	-0.82	-24.57	-13.37
1120	-9.82	-10.23	-0.41	-23.79	-13.97
1140	-10.34	-11.38	-1.04	-25.66	-15.32
1160	-16.61	-18.39	-1.78	-30.53	-13.92
1180	-12.91	-15.64	-2.73	-31.76	-18.85
1200	-12.51	-12.54	-0.03	-29.85	-17.34
1220	-13.73	-15.79	-2.06	-36.92	-23.19
1240	-12.58	-14.12	-1.54	-30.66	-18.08
1260	-12.19	-12.52	-0.33	-28.16	-15.97
1280	-14.75	-15.85	-1.1	-32.22	-17.47
1300	-17.61	-18.48	-0.87	-29.34	-11.73

5. 1 GHz – 1.3 GHz vertical polarisation:

frequency in MHz	free-space loss in dB	standard iso in dB	Δ in dB	heat – iso in dB	Δ in dB
1000	-7.94	-8.45	-0.51	-32.38	-24.44
1020	-7.75	-9.05	-1.3	-22.53	-14.78
1040	-9.94	-11.85	-1.91	-21.4	-11.46
1060	-12.42	-14.69	-2.27	-20.73	-8.31
1080	-14.48	-17.56	-3.08	-20.53	-6.05
1100	-19.7	-28.39	-8.69	-20.21	-0.51
1120	-15.76	-19.02	-3.26	-20.64	-4.88
1140	-14.63	-16.03	-1.4	-23.94	-9.31
1160	-15.3	-16.83	1.53	-28.68	-13.38
1180	-8.99	-9.33	-0.34	-23.62	-14.63
1200	-8.53	-9.43	-0.9	-27.09	-18.56
1220	-8.71	-9.94	-1.23	-26.87	-18.16
1240	-8.57	-9.18	-0.61	-30.53	-21.96
1260	-8.25	-9.12	-0.87	-29.68	-21.43
1280	-11.27	-12.57	-1.3	-32.64	-21.37
1300	-14.07	-14.97	-0.9	-32.44	-18.37

6. 2.2 GHz – 3 GHz horizontal polarisation (standard isolation window):

frequency in GHz	free-space loss in dB	standard iso in dB	Δ in dB
2.2	-9.76	-10.69	-0.93
2.3	-11.49	-12.94	-1.45
2.4	-12.73	-13.58	-0.85
2.5	14.63	-15.39	-0.76
2.6	-14.02	-14.71	-0.69
2.7	-12.14	-12.34	-0.2
2.8	-8.49	-9.92	-1.43
2.9	-6.41	-7.07	-0.66
3	-9.5	-10.18	-0.68

7. 2.2 GHz – 3 GHz vertical polarisation (standard isolation window):

frequency in GHz	free-space loss in dB	standard iso in dB	Δ in dB
2.2	-10.9	-12.35	-1.45
2.3	-11.87	-12.32	-0.45
2.4	-15.17	-15.94	-0.77
2.5	-15.1	-16.02	-0.92
2.6	-14.14	-14.78	-0.64
2.7	-12.58	-12.97	-0.39
2.8	-9.46	-10.45	-0.99
2.9	-7.12	-8.29	-1.17
3	-9.15	-9.9	-0.75

3. 4/0 MHz – 862 MHz (heat isolation window):

8. 2.2 GHz – 3 GHz horizontal polarisation (heat isolation window):

frequency in GHz	free-space loss in dB	heat – iso in dB	Δ in dB
2.2	-8.02	-24.75	-16.73
2.3	-7.35	-24.42	-17.07
2.4	-11.97	-27.46	-15.49
2.5	-9.96	-31.67	-21.71
2.6	-9.28	-28.07	-18.79
2.7	-8.07	-26.18	-18.11
2.8	-5.15	-25.13	-19.98
2.9	-8.5	-28.8	-20.3
3	-8.15	-40.1	-31.95

9. 2.2 GHz – 3 GHz vertical polarisation (heat isolation window):

frequency in GHz	free-space loss in dB	heat – iso in dB	Δ in dB
2.2	-5.98	-24.95	-18.97
2.3	-9.27	-34.31	-25.04
2.4	-10.24	-34.76	-24.52
2.5	-10.58	-33.82	-23.24
2.6	-10.62	-30.81	-20.19
2.7	-7.82	-35.46	-27.64
2.8	-5.96	-40.04	-34.08
2.9	-8.39	-33.81	-25.42
3	-10.12	-31.42	-21.3

10. 8 GHz – 12.5 GHz horizontal polarisation

frequency in GHz	free-space loss in dB	standard iso in dB	Δ in dB	heat – iso in dB	Δ in dB
8	-5.05	-17.03	-11.98	-39.37	-34.32
8.1	-4.98	-15.48	-10.5	-36.34	-31.36
8.2	-6.91	-17.84	-10.93	-42.84	-35.93
8.3	-7.08	-17.91	-10.83	-47.9	-40.82
8.4	-7.7	-16.77	-9.07	-41.85	-34.15
8.5	-5.8	-16.06	-10.26	-43.63	-37.83
8.6	-5.56	-13.93	-8.37	-36.68	-31.12
8.7	-4.1	-12.54	-8.44	-42.67	-38.57
8.8	-3.86	-12.23	-8.37	-35.73	-31.87
8.9	-4.68	-13.35	-8.67	-41.34	-36.66
9	-5.71	-11.58	-5.87	-39.95	-34.24
9.1	-5.58	-12.58	-7	-38.24	-32.66
9.2	-6.85	-10.47	-3.62	-41.29	-34.44
9.3	-4.14	-11.66	-7.52	-34.55	-30.41
9.4	-5.9	-10.21	-4.31	-39.47	-33.57
9.5	-3.68	-10.79	-7.11	-35.04	-31.36
9.6	-6.05	-10.02	-3.97	-34.74	-28.69
9.7	-4.67	-9.54	-4.87	-36.64	-31.97
9.8	C.68	-10.48	-3.9	-36.44	-29.86
9.9	-9.24	-12.68	-3.44	-38.77	-29.53
10	-8.4	-11.88	-3.48	-34.19	-25.79
10.1	-7.49	-8.71	-1.22	-30.56	-23.07
10.2	-8.42	-10.36	-1.94	-34.03	-25.61
10.3	-4.79	-6.08	-1.29	-33.44	-28.65
10.4	-9.43	-10.59	-1.16	-36.25	-26.82
10.5	-8.06	-9.35	-1.29	-36.31	-28.25
10.6	-9.13	-10.43	-1.3	-35.74	-26.61
10.7	-9.6	-10.55	-0.95	-36.86	-27.26
10.8	-8.87	-10.18	-1.31	-42.73	-33.86
10.9	-10.47	-10.93	-0.46	-38.79	-28.32
11	-7.85	-8.84	-0.99	-40.54	-32.69
11.1	-6.79	-7.18	-0.39	-35.03	-28.24
11.2	-7.62	-8.32	-0.7	-34.5	-26.88
11.3	-7.67	-9.22	-1.55	-38.87	-31.2
11.4	-8.94	-10.75	-1.81	-39.9	-30.96
11.5	-8.79	-10.67	-1.88	-38.49	-29.7
11.6	-9.78	-11.79	-2.01	-36.25	-26.47
11.7	-8.32	-10	-1.68	-36.25	-27.93
11.8	-7.01	-9.03	-2.02	-37.95	-30.94
11.9	-8.81	-10.98	-2.17	-36.49	-27.68
12	-6.87	-8.8	-1.93	-36.65	-29.78
12.1	-7.17	-9.78	-2.61	-36.02	-28.85
12.2	-7.58	-10.17	-2.59	-37.14	-29.56
12.3	-9.68	-11.87	-2.19	-38.23	-28.55
12.4	-10.41	-12.96	-2.55	-38.61	-28.2
12.5	-9.89	-12.23	-2.34	-35.96	-26.07

11. 8 GHz – 12.5 GHz vertical polarisation

frequency in GHz	free-space loss in dB	standard iso in dB	Δ in dB	heat – iso in dB	Δ in dB
8	-4.19	-14.45	-10.26	-38.02	-33.83
8.1	-5.79	-15.13	-9.34	-41.56	-35.77
8.2	-6.84	-16.83	-9.99	-41.74	-34.9
8.3	-6.86	-15.49	-8.63	-38.18	-31.32
8.4	-5.98	-15.95	-9.97	-39.37	-33.39
8.5	-5.71	-14.93	-9.22	-37.69	-31.98
8.6	-4.63	-13.49	-8.86	-39.86	-35.23
8.7	-3.6	-12.88	-9.28	-37.21	-33.61
8.8	-3.26	-12.14	-8.88	-37.54	-34.28
8.9	-4.15	-12.2	-8.05	-40.69	-36.54
9	-5.62	-13.28	-7.66	-38.18	-32.56
9.1	-5.36	-12.29	-6.93	-40.07	-34.71
9.2	-5.8	-13.32	-7.52	-39.56	-33.76
9.3	-5.11	-11.12	-6.01	-37.47	-32.36
9.4	-4.43	-11.05	-6.62	-37.86	-33.43
9.5	-3.92	-9.49	-5.57	-35.88	-31.96
9.6	-5.54	-11.02	-5.48	-37.16	-31.62
9.7	-5.79	-11.34	-5.55	-37.62	-31.83
9.8	-6.94	-11.38	-4.44	-40.08	-33.14
9.9	-8.75	-12.91	-4.16	-42.52	-33.77
10	-8.28	-11.72	-3.44	-41.81	-33.53
10.1	-8.18	-11.02	-2.84	-45.17	-36.99
10.2	-6.9	-9.88	-2.98	-37.39	-30.49
10.3	-7.02	-8.64	-1.62	-38.77	-31.75
10.4	-6.62	-8.11	-1.49	-37.82	-31.2
10.5	-9.25	-10.19	-0.94	-44.32	-35.07
10.6	-8.92	-9.67	-0.75	-42.24	-33.32
10.7	-8.93	-9.37	-0.44	-39.95	-31.02
10.8	-9.53	-10.03	-0.5	-37.61	-28.08
10.9	-8.92	-9.25	-0.33	-37.55	-28.63
11	-8.45	-9.02	-0.57	-36.06	-27.61
11.1	-7.83	-8.55	-0.75	-36.28	-28.45
11.2	-7.26	-7.83	-0.57	-35.27	-28.01
11.3	-7.27	-8.2	-0.93	-36.54	-29.27
11.4	-9.64	-10.52	-0.88	-35.53	-25.89
11.5	-9.51	-10.48	-0.97	-36.25	-26.74
11.6	-10.04	-11.51	-1.47	-36.35	-26.31
11.7	-8.31	-9.39	-1.08	-36.29	-27.98
11.8	-8.39	-9.81	-1.42	-35.77	-27.38
11.9	-6.86	-8.43	-1.57	-35.04	-28.18
12	-7.81	-9.63	-1.82	-35.87	-28.06
12.1	-7.66	-9.69	-2.03	-34.78	-27.12
12.2	-8.06	-9.75	-1.69	-37.18	-29.12
12.3	-10.02	-11.96	-1.94	-38.56	-28.54
12.4	-8.9	-11.42	-2.52	-40.78	-31.88
12.5	-9.36	-11.67	-2.31	-39.6	-30.24

Simulation results to 5.1:

vertical polarisation:

$\varepsilon_1 = 1 ; \varepsilon_2 = 4$		$\varepsilon_1 = 1 ; \varepsilon_2 = 6$		$\varepsilon_1 = 1 ; \varepsilon_2 = 8$		$\varepsilon_1 = 1 ; \varepsilon_2 = 10$	
incident angle	reflection coeff.	incident angle	reflection coeff.	incident angle	reflection coeff.	incident angle	reflection coeff.
0	1	0	1	0	1	0	1
5	0.67	5	0.62	5	0.58	5	0.55
10	0.43	10	0.37	10	0.31	10	0.27
15	0.26	15	0.18	15	0.12	15	0.08
20	0.13	20	0.05	20	0.01	17	0
25	0.03	23	0	25	0.12	20	0.06
27	0	25	0.05	30	0.2	25	0.17
30	0.05	30	0.13	35	0.26	30	0.24
35	0.11	35	0.2	40	0.31	35	0.31
40	0.16	40	0.25	45	0.35	40	0.35
45	0.2	45	0.29	50	0.38	45	0.39
50	0.24	50	0.32	55	0.41	50	0.43
55	0.26	55	0.35	60	0.43	55	0.45
60	0.28	60	0.37	65	0.44	60	0.47
65	0.3	65	0.39	70	0.46	65	0.49
70	0.31	70	0.4	75	0.47	70	0.5
75	0.32	75	0.41	80	0.47	75	0.51
80	0.33	80	0.42	85	0.48	80	0.52
85	0.33	85	0.42	90	0.48	85	0.52
90	0.33	90	0.42			90	0.52

horizontal polarisation:

$\varepsilon_1 = 1 ; \varepsilon_2 = 4$		$\varepsilon_1 = 1 ; \varepsilon_2 = 6$		$\varepsilon_1 = 1 , \varepsilon_2 = 8$		$\varepsilon_1 = 1 ; \varepsilon_2 = 10$	
incident angle	reflection coeff.	incident angle	reflection coeff.	incident angle	reflection coeff.	incident angle	reflection coeff.
0	1	0	1	0	1	0	1
5	0.9	5	0.93	5	0.94	5	0.94
10	0.82	10	0.86	10	0.88	10	0.89
15	0.74	15	0.79	15	0.82	15	0.84
20	0.68	20	0.74	20	0.77	20	0.8
25	0.62	25	0.69	25	0.73	25	0.76
30	0.57	30	0.64	30	0.69	30	0.72
35	0.52	35	0.6	35	0.65	35	0.68
40	0.48	40	0.57	40	0.62	40	0.65
45	0.45	45	0.54	45	0.59	45	0.63
50	0.42	50	0.51	50	0.56	50	0.6
55	0.4	55	0.49	55	0.54	55	0.58
60	0.38	60	0.47	60	0.53	60	0.57
65	0.37	65	0.45	65	0.51	65	0.55
70	0.35	70	0.44	70	0.5	70	0.54
75	0.34	75	0.43	75	0.49	75	0.53
80	0.34	80	0.43	80	0.48	80	0.52
85	0.33	85	0.42	85	0.48	85	0.52
90	0.33	90	0.42	90	0.48	90	0.52

Simulation results to 6.3:

permittivity	inc. angle = 0°	inc. angle = 5°	inc. a ngle = 10°	inc. angle = 20°
	losses in dB	losses in dB	losses in dB	losses in dB
1	0	0	0	0
2	-0.52	-0.53	-0.54	-0.62
3	-1.3	-1.31	-1.34	-1.5
4	-2.05	-2.07	-2.12	-2.33
5	-2.74	-2.76	-2.82	-3.09
6	-3.38	-3.4	-3.47	-3.77
7	-3.96	-3.99	-4.06	-4.39
8	-4.5	-4.53	-4.61	-4.96
9	-5	-5.03	-5.12	-5.49
10	-5.47	-5.5	-5.59	-5.98
11	-5.9	-5.94	-6.03	-6.44
12	-6.32	-6.35	-6.45	-6.87
13	-6.7	-6.74	-6.84	-7.28
14	-7.07	-7.11	-7.22	-7.66
15	-7.42	-7.46	-7.57	-8.03
16	-7.76	-7.79	-7.91	-8.37
17	-8.07	-8.11	-8.23	-8.71
18	-8.38	-8.42	-8.53	-9.02
19	-8.67	-8.71	-8.83	-9.32
20	-8.95	-8.99	-9.11	-9.61
21	-9.22	-9.26	-9.39	-9.89
22	-9.48	-9.52	-9.65	-10.16
23	-9.73	-9.78	-9.9	-10.42
24	-9.98	-10.02	-10.15	-10.67
25	-10.21	-10.26	-10.38	-10.91
26	-10.44	-10.48	-10.61	-11.15
27	-10.66	-10.71	-10.84	-11.38
28	-10.88	-10.92	-11.05	-11.6
29	-11.09	-11.13	-11.26	-11.81
30	-11.29	-11.33	-11.47	-12.02

A4: References

[1] Theodore S. Rappaport, "*Wireless Communications, Principles and Practice*", Second Edition, Prentice Hall PTR, 2002

[2] Douglas C. Giancoli, "*Physics for scientists and engineers with modern physics*", Second Edition, Prentice-Hall International, Inc. Englewood Cliffs, New Jersey, 1988

[3] Constantine A. Balanis," *Advanced Engineering Electromagnetics*", John Wiley & Sons, Inc., 1989

[4] Jørgen Bach Andersen, Theodore S. Rappaport, Susumu Yoshida, "*Propagation Measurements and Models for Wireless Communications Channels*", IEEE Communications Magazine, Jan 1995

[5] Harris Benson, "*University Physics*", Revised Edition, John Wiley & Sons, Inc., 1996

[6] John Avison, "*The World of Physics*", Second Edition, Thomas Nelson and Sons Ltd, 1989

[7] Richard Feynman, "*The Feynman Lectures on Physics: Commemorative Issue*", Volume 2, Addison-Wesley Publishing Co., Inc., Reading, Massachusetts, 1989

[8] Clayton R. Paul, Keith W. Whites, Syed A. Nasar, "*Introduction to electromagnetic fields*", Third Edition, McGraw-Hill

[9] Prof. Dipl.-Ing. H. Dölecke, "*Radartechnik*", Vorlesungsskript, Fachhochschule Hannover,1999

[10] Antonio Fischer de Toledo, Adel M. D. Turkmani, J. David Parsone, "*Estimating Coverage of Radio Transmission into and within Buildings at 900, 1800, and 2300 MHz* ", IEEE Personal Communications, April 1998

[11] http://de.wikipedia.org/wiki/Glas

[12] H. Rawson, "*Glass and its history of service*", IEE Proceedings, July 1988

[13] Glas, Lehrstuhl Baukonstruktion 2, RWTH Aachen

[14] Saint-Gobain Glass "*Handbuch-Toleranzen 2.Auflage 2006*", 2006

[15] Robert Wilson," *Propagation Losses Through Common Building Materials*", University of Southern California, August 2002

[16] United States Patent, Patent Number : 5,147,694, "*Electromagnetic Shielding Panel*", Sep. 1992

[17] United States Patent, Patent Number : 5,776,612, "*Window that transmits light energy and selectively absorbs microwave energy*", July 1998

[18] Björn Widenberg, José Víctor Rodríguez Rodríguez, "*Design of Energy Saving Windows with High Transmission at 900 MHz and 1800 MHz*", Department of Electroscience, Electromagnetic Theory, Lund Institute of Technology, Sweden, 2002

[19] Hirai, Yokota, "*Electro-Magnetic Shielding Glass of frequency selective surfaces*", Kajima Technical Research Institute Chofu-shi, Tokyo, Japan

[20] GlassLock, SpyGuard™

[21] James Baker-Jarvis, "*Dielectric Characterization of Low-Loss Materials*", IEEE Transactions on Dielectrics and Electric Insulation, August 1998

[22] P. Hui, "*Minimization of Microwave Reflections from EC-Coated Glass Slabs*", IEEE Microwave and Guided Wave Letters, Nov. 1998

[23] Cuinas, García Sánchez," *Measuring, Modeling, and Characterizing of Indoor Radio Channel at 5.8 GHz*", IEEE Transactions on Vehicular Technology, March 2001

[24] John J. Holmes, Constantine A. Balanis, "*Refraction of a Uniform Plane Wave Incident on a Plane Boundary Between Two Lossy Media*", IEEE Transactions on Antennas and Propagation, Sep. 1978

[25] Daniel Peña, Rodolfo Feick, "*Measurement and Modeling of Propagation Losses in Brick and Concrete Walls for the 900 MHz-Band*", IEEE Transactions on Antennas and Propagation, Jan 2003

[26] NIST Construction Automation Program Report No. 3,"*Electromagnetic Signal Attenuation in Construction Materials*", United States Department of Commerce, Gaithersburg, Maryland

[27] Helena Bülow-Hübe, "*Energy-Efficient Windows Systems*", Doctoral Dissertation, Department of Construction and Architecture, Lund Institute of Technology, Sweden, 2001

[28] Pilkington Basisgläser 2006

[29] Agilent, "*Basics of Measuring the Dielectric Properties of Materials*"